T0269274

Spectral Geometry of Shapes

Principles and Applications

Computer Vision and Pattern Recognition Series

Series Editors

Also in the Series:

Lin and Zhang, Low-Rank Models in Visual Analysis: Theories, Algorithms and Applications, 2017, ISBN: 9780128127315
Zheng et al., Statistical Shape and Deformation Analysis: Methods, Implementation and Applications, 2017, ISBN: 9780128104934
De Marsico et al., Human Recognition in Unconstrained Environments: Using Computer Vision, Pattern Recognition and Machine Learning Methods for Biometrics, 2017, ISBN: 9780081007051
Saha et al., Skeletonization: Theory, Methods and Applications, 2017, ISBN: 9780081012918
Murino et al., Group and Crowd Behavior for Computer Vision, 2017, ISBN: 9780128092767
Leo and Farinella, Computer Vision for Assistive Healthcare, 2018, ISBN: 9780128134450
Alameda-Pineda et al., Multimodal Behavior Analysis in the Wild, 2019, Advances and Challenges, ISBN: 9780128146019
Wang et al., Deep Learning Through Sparse and Low-Rank Modeling, 2019, ISBN: 9780128136591
Ying Yang, Multimodal Scene Understanding: Algorithms, Applications and Deep Learning, 2019, ISBN: 9780128173589

Spectral Geometry of Shapes

Principles and Applications

Jing Hua
Wayne State University, Department of Computer Science,
Detroit, MI, United States

Jiaxi Hu
Google Inc., Mountain View, CA, United States

Zichun Zhong
Wayne State University, Department of Computer Science,
Detroit, MI, United States

Academic Press is an imprint of Elsevier
125 London Wall, London EC2Y 5AS, United Kingdom
525 B Street, Suite 1650, San Diego, CA 92101, United States
50 Hampshire Street, 5th Floor, Cambridge, MA 02139, United States
The Boulevard, Langford Lane, Kidlington, Oxford OX5 1GB, United Kingdom

Library of Congress Cataloging-in-Publication Data
A catalog record for this book is available from the Library of Congress

British Library Cataloguing-in-Publication Data
A catalogue record for this book is available from the British Library

ISBN: 978-0-12-813842-7

For information on all Academic Press publications
visit our website at https://www.elsevier.com/books-and-journals

Publisher: Mara Conner
Acquisition Editor: Tim Pitts
Editorial Project Manager: Joshua Mearns
Production Project Manager: Kamesh Ramajogi
Designer: Christian J. Bilbow

Typeset by VTeX

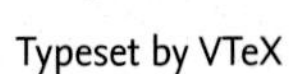

Contents

About the authors

Jing Hua is Professor of Computer Science at Wayne State University. He received his Ph.D. degree (2004) in Computer Science from the State University of New York at Stony Brook. He also received his M.S. degree (1999) in Pattern Recognition and Artificial Intelligence from the Institute of Automation, Chinese Academy of Sciences in Beijing, China, and his B.S. degree (1996) in Electrical Engineering from the Huazhong University of Science and Technology in Wuhan, China. His research interests include Geometric Modeling of Shapes, Computer Graphics, Visualization, Image Analysis and Informatics, Computer Vision, and so on. He has authored over 150 papers in these research fields. He received the Gaheon Award for the Best Paper of International Journal of CAD/CAM in 2009, the Best Paper Award at ACM Solid Modeling 2004, the WSU Faculty Research Award in 2005, the College of Liberal Arts and Sciences Excellence in Teaching Award in 2008, the K. C. Wong Research Award in 2010, and the Best Demo Awards at GENI Engineering Conference 21 (2014) and 23 (2015). His research is funded by the National Science Foundation, National Institutes of Health, Michigan Technology Tri-Corridor, Michigan Economic Development Corporation, Ford Motor Company, and WSU. He serves as an Editorial Board Member for many journals and a Program Committee Member for many international conferences.

Jiaxi Hu is a software engineer at Google LLC, Mountain View, California. He received his B.S. degree (2003) and M.S. degree (2005) from the Huazhong University of Science and Technology, Wuhan, China, and Ph.D. degree (2014) from Wayne State University, Detroit, Michigan, before he joined Google LLC in 2014. After a decade of research of Computer Graphics in the university, he brought his expertise to the world through Google web services. He worked on Imagery Viewer, which is a backend rendering engine to serve a broad range of Google products, such as Google Map, Street View, local search, and so on, and third-party partners. After that, he found the new challenge in bringing Google Assistant to Next Billion Users.

Zichun Zhong is Assistant Professor of Computer Science at Wayne State University. He received his Ph.D. degree (2014) in Computer Science at The University of Texas at Dallas. He received B.S. degree (2006) and M.S. degree (2009) in Computer Science from the University of Electronic Science and Technology of China (UESTC) in Chengdu, China. He was a Postdoctoral Fellow (2014–2015) in Department of Radiation Oncology at UT Southwestern Medical Center at Dallas. His research interests include Computer Graphics, Geometric Modeling (specifically surface and volume mesh generations), Medical Image Processing (specifically deformable image registration, 3D/4D image reconstruction), Visualization, Virtual Reality and Augmented Reality, and GPU Algorithms. He has published about 50 conference and journal papers in these research fields, including ACM SIGGRAPH, ACM Transactions on Graphics, IEEE Visualization, IEEE Transactions on Visualization and Computer Graphics, CVPR, ICCV, MICCAI, and so on. He received the National Science Foundation (NSF) CAREER Award in 2019, Faculty Research Excellence Award at WSU in 2019, and Certificate of Academic Excellence at UTD in 2013. He serves as a program committee member for many international conferences and a reviewer for many top conferences and journals in his research field.

Preface

Shape analysis is a fundamental problem in many research fields such as computer graphics, computer vision, image processing, robotics, and so on. Due to the rapid advancement of 3D acquisition technologies, massive 3D shapes are being produced using laser scanning, structure light cameras, Microsoft Kinect and Intel RealSense cameras, as well as many medical scanners, such as MRI, CT, and 3D Ultrasound. The increasing 3D shape data demand a variety of shape analysis methods. Considering shape as general data, there exist many highly demanded analyses, for example, matching, indexing, retrieval, registration, and mapping. On top of these ones, high-level understanding is also desired, including 4D time-varying shape motion, shape evolution, and so on. Nevertheless, 3D shape analysis based on Euclidean transformations suffers many variance problems. This book presents unique shape analysis approaches based on shape spectrum in differential geometry. The shape spectra contain the intrinsic geometry information of a shape and reveal the relationship across shapes; therefore they are able to facilitate all the aforementioned 3D shape analyses in a single uniform framework, handling both isometric and nonisometric variations and supporting deep learning of shape geometry. The book also demonstrates several exemplar applications of using the shape spectrum-based methods for solving real-world problems.

This book presents latest advancement in spectral geometric processing for 3D shape analysis applications, such as shape classification, shape matching, medical imaging, and so on. It provides intuitive links between fundamental geometry theories and real-world applications, bridging the gap between the theory and practice; it describes new theoretical breakthroughs in applying spectral methods for nonisometric motion analysis; it presents readers the insights to develop spectral geometry-based approaches for 3D shape analysis; it introduces deep learning of spectral geometry using convolutional neural networks and its application to shape classification and retrieval in a large-scale 3D shape database.

Jing Hua, Jiaxi Hu, and Zichun Zhong
May 1, 2019

Acknowledgments

We thank Tim Pitts and Joshua Mearns for guidance throughout the publishing process.

Jing Hua, Jiaxi Hu, and Zichun Zhong
May 1, 2019

Chapter 1

Introduction

Contents

1.1 Overview of shape analysis and applications

Shape analysis is a fundamental problem in many research fields such as computer graphics, computer vision, image processing, robotics, and so on. Computer scientists and engineers consider shape as an attribute to describe an object [89,61]. In the past decade, massive 3D shapes are produced with advanced technologies. Traditional computer-aided design (CAD) creates a lot of manufacturing 3D models [114]. Laser scanning generates 3D point clouds or surface meshes [78]. Similarly, structure light camera generates depth images. For example, Kinect from Microsoft reduced the cost of this technique and made it available for daily use. MRI or CT scanning produces intensity-based volume data. Although 3D shape data are presented in various formats, those formats can be converted to each other. For example, surfaces can be reconstructed from point clouds or extracted from the isovalues of the intensity-based volume. In this work, we focus on the shapes represented with two differentiable manifold boundary surfaces. In practice, these manifolds are discretized into triangle or tetrahedron meshes. The increasing 3D shape data demand a variety of shape analysis methods. Considering shapes as general data, there exist basic analyses, for example, matching, indexing, retrieval, registration, and mapping. On top of these basic ones, high-level understanding is also desired, including pose analysis, 4D time-varying motion, and so on.

Traditional shape analysis starts from the original spatial properties of shapes, for example, curvature, diameter, and geodesic distance. There are also more advanced shape representations. For example, moments describe a shape with a set of integrations of different orders [26,123,29,101]. An extended Gaussian image is built with the orientation and area information of a convex polygon mesh for shape representation. This extended Gaussian image can uniquely describe a convex mesh [41,132]. Shape distributions measure properties based on distance, angle, area, and volume measurements between random surface points. The similarity between the objects is measured by a pseudometric that measures distances between distributions [105,103]. Geometric hashing represents a shape with a set of its local interest features (points, lines, or other

Spectral Geometry of Shapes. https://doi.org/10.1016/B978-0-12-813842-7.00009-7

suitable features) [73,77,147,35]. A shape can be described by another kind of data structure, such as vectors or graphs [137,50,87,141,125]. Graph-based approaches analyze a 3D shape by transforming it to a graph, such as a B-Rep graph, a Reeb graph, and a skeletal graph, and convert shape analysis into graph problems [27,28,39,9,20,9]. Spherical harmonics represents a shape by a 2D histogram of radius and frequency [141,125,59,60,102]. It decomposes a model into a collection of spherical functions on the concentric spheres, then calculates the Fourier transforms of these spherical functions. Shape histograms [2] describe a shape by the partitions of the 3D space. The 3D space can be decomposed into disjoint cells in different ways. These traditional shape analysis approaches are often challenged by Euclidean transformations, irregular mesh samplings, and nonlinear deformations.

Another category of shape analysis is based on geometric mappings. A shape mapping is a powerful tool to reduce the complexity of arbitrary manifolds onto canonical domains such as unit cube or sphere, where regular analysis, for example, image-based processing, can be applied directly [72]. Among this category, functional methods typically start with defining certain penalty functions such that the minima are assumed at desired results. The mapping is then achieved using optimization methods. A conformal mapping provides a unique mapping by preserving local angle geometries. Conformal methods possess several unique advantages, for example, exact angle preserving, guarantee of solution existence, an efficient algorithm, and a rich continuous theory in parallel. At the mean time, a conformal mapping introduces large area distortions. To reduce the area distortions, some additional processes are applied. Gu and Yau [34] punctured small holes at the tip of long appendages. Cone singularities were introduced with nonvanishing Gaussian curvature in [62,7]. Surface cuts were repeatedly augmented according to the geometric stretches generated through the course of tentative parameterizations [33]. Zou et al. [159] presented a practical method to compute a group of analytic global 2D area-preserving mappings mathematically with *Lie advection*. A manifold cannot be mapped to another domain without any distortion. Thus different mapping methods have been proposed to preserve certain local geometries [33,80]. That is the dilemma of the mapping-based approaches for shape analysis.

The spectrum-based approach is inspired by the Fourier transform in signal processing, where the time variant signals can be projected onto functional bases. An early shape spectrum work is applied on graphs [96,97,93]. Considering discrete meshes as graphs, a Laplacian matrix is defined on vertices and connections, and weights may be also applied. The eigenvalues are defined as the spectra of graphs, and the eigenfunctions are the orthogonal bases. This spectrum has a lot of similarities with the Fourier transform. The graphs are then projected onto those bases and analyzed in the spectral domain. Karni and Gotsman [58] used the projections of geometry on the eigenfunctions for mesh compression and smoothing. Jain and Zhang [51] extended it for shape registration in the spectral domain. The Laplace spectrum focuses on the connection of graph instead of the intrinsic geometry of the manifolds. Only using the connectivity of the graph may lead to highly distorted mappings [158].

From the view of computational geometry, the geometry of the shape can be represented by a differentiable manifold. Reuter [116,115,119,118] defined the shape spectrum as the family of eigenvalues of the Laplace–Beltrami operator on a manifold. Nearly at the same time, Reuter [116] and Lévy [79] found out the shape spectrum as the family of eigenvalues of the Laplace–Beltrami operator on a manifold and used it as a global shape descriptor. Without further notation, the shape spectrum in this work refers to the Laplace–Beltrami spectrum. Laplace–Beltrami eigenfunctions are also tools to understand the geometry. Rustamov [121] proposed a modified shape distribution based on eigenfunctions and eigenvalues. It is proved that the shape spectrum is invariant to spatial translation, rotation, scaling, and isometric deformations. The spectrum is also stable to different triangulations and near-isometric deformations. The shape spectrum describes the similarities among shapes. The shape spectrum has a lot of great properties for shape analysis. It is invariant to Euclidean transformations and isometric deformations. On discrete triangle meshes, the spectrum is invariant to different triangulations. It carries the intrinsic geometry of the manifold behind various representations. However, by definition the spectrum depends on the global geometry. It changes a lot while the geometry changes. Analysis and experiments show that it is stable among near-isometric deformations and minor noises. A greater nonisometric deformation breaks the connections among different objects. This often restricts the shape spectrum to a global shape descriptor or the same object analysis.

To solve those research problems, we present the shape analysis approaches by employing the shape spectrum in differential geometry. The shape spectra contain the intrinsic geometry information of the original shape and reveal the relationship across shapes. This book is organized as follows:

- Chapter 2 presents the background of shape analysis using spectral geometry. It reviews the definition of the shape spectrum and numerical computations.
- Chapter 3 presents a method to extract salient spectral geometric features in the spectral domain derived from the Laplace–Beltrami operator, which is invariant to Euclidean transformations and isometric deformations. Describing and matching shapes with their salient features also conform to the procedure of "coarse-to-fine" multiresolution analysis. The features are extracted with local "frequencies" identified, which imply spatial scales of local support regions defining the features, that is, each salient feature finds its local support. The salient spectral features are very stable and distinctive. The shape representation built upon them may achieve a higher level of shape description. It can be easily applied to tasks such as shape matching and shape retrieval.
- Chapter 4 presents a registration-free shape motion analysis method based on the Laplace–Beltrami spectral domain. By transferring the shapes from the spatial domain to the spectral domain all the Euclidean transformations and isometric deformations are filtered out. In the spectral domain, different poses

are understood based on meaningful parts. Surface mesh vertices belonging to the same semantic part on different pose surfaces will be mapped to the same coordinates in the geometry spectral domain while they carry different spatial properties under different poses. The analysis of the spatial property variation in the geometry spectral domain is able to quantify the geometry behavior of every point during the pose changes and, consequently, to classify a point to a rigid part or an articulated part in the spectral domain. The shape is then decomposed into parts with different geometric semantics. The skeleton can be generated automatically based on eigenfunctions of the shape. The behavior of the skeleton is constrained by the surface properties and classified surface semantics, which also represents the semantics of that skeleton.

- Chapter 5 presents an analytic method to align spectrum among different shapes with nonisometric deformations. It is proved that eigenvalues are continuous functions of scalar factors applied on the conformal metric. The derivatives of the eigenvalues are analytically expressed by those of the scalar field defined on the original manifold. The discrete counterpart on the triangle meshes also follows the same behavior. In this case the scalar field turns into a scale vector on the mesh whose value is sampled at each vertex. More specifically, the analytic expression between the eigenvalues and the scale vector can be reformed into a linear system. With smoothness and local bound constraints, the linear system is consequently solvable as a quadratic problem. Then the spectra can be controlled with local scale vectors. The approach closes the gap that shape spectrum is not invariant to nonisometric deformations. The scale field is the solution of a quadratic problem. By applying such a scale field the spectra are aligned among shapes with nonisometric deformations. On triangle meshes the scalar scale factors are represented with a scale vector defined on each vertex.
- Chapter 6 presents a novel surface registration technique using the spectrum of the shapes, which can facilitate accurate localization and visualization of nonisometric deformations of the surfaces. To register two surfaces, we map both eigenvalues and eigenvectors of the Laplace–Beltrami operator of the shapes through optimizing an energy function. The function is defined by the integration of a smoothness term to align the eigenvalues and a distance term between the eigenvectors at feature points to align the eigenvectors. The feature points are generated using the static points of certain eigenvectors of the surfaces. By using both the eigenvalues and the eigenvectors on these feature points the computational efficiency is improved considerably without losing the accuracy in comparison to the approaches that use the eigenvectors for all vertices. In our technique the variation of the shape is expressed using a scale function defined at each vertex. Consequently, the total energy function to align the two given surfaces can be defined using the linear interpolation of the scale function derivatives. Through the optimization of the energy function, the scale function can be solved, and the alignment is achieved. After the alignment, the eigenvectors can be employed to calculate the point-to-

point correspondence of the surfaces. Therefore the proposed method can accurately define the displacement of the vertices.

- Chapter 7 describes the main mathematical ideas behind geometric deep learning and provides development details for several applications in shape analysis and synthesis, especially in spectral geometry. It provides a clear picture of the key concepts and techniques, and illustrates the related applications. We also aim to provide practical implementation details for the methods presented in these works.
- Chapter 8 gives a summary of the book with the major contributions and potential future work.

Chapter 2

Background

Contents

In this book a shape is represented by a differentiable manifold in the computational geometry. On such a manifold the differential operator is defined based on the local geometry. Within one shape the operator contains intrinsic geometry information. The differential operator will introduce the shape spectrum on the original manifolds. In this chapter, we are going to briefly review the definition of the shape spectrum and numerical computations.

2.1 Laplace shape spectrum

In this section, we review the theory of Laplacian spectrum and describe how to compute it on a triangle mesh.

Let $f \in C^2$ be a real function on a Riemannian manifold M. The Laplace–Beltrami operator Δ is defined as

$$\Delta f = \nabla \cdot (\nabla f), \tag{2.1}$$

where ∇f is the gradient of f, and $\nabla \cdot$ is the divergence on the manifold M [18]. The Laplace–Beltrami operator is a linear differential operator and can be calculated in local coordinates. Let ψ be a local parameterization of a submanifold of M such that $\psi : R^n \rightarrow R^{n+k}$, $g_{ij} =< \partial_i \psi, \partial_j \psi >$, $G = (g_{ij})$, $W = \sqrt{\det G}$, and $(g^{ij}) = G^{-1}$, where $i, j = 1, 2, \ldots, n$, $<, >$ is the dot product, and det is the determinant. The Laplace–Beltrami operator then is defined on the submanifold as $\Delta f = \frac{1}{W} \sum_{i,j} \partial_i (g^{ij} W \partial_j f)$. If $M \subset R^2$, then the Laplace–Beltrami operator reduces to the Laplacian:

$$\Delta f = \frac{\partial^2 f}{(\partial x)^2} + \frac{\partial^2 f}{(\partial y)^2}. \tag{2.2}$$

Consider the Laplacian eigenvalue equation

$$\Delta f = -\lambda f, \tag{2.3}$$

Spectral Geometry of Shapes. https://doi.org/10.1016/B978-0-12-813842-7.00010-3

where λ is a real scalar. The solution is a family of nonnegative scalars $\{\lambda_i\}$ and the corresponding real functions $\{f_i\}$ for $i = 0, 1, 2, \ldots$. The spectrum is defined as the eigenvalues arranged increasingly as $0 \leq \lambda_0 \leq \lambda_1 \leq \lambda_2 \leq \cdots < +\infty$. In the case of a closed manifold or an open manifold with Neumann boundary condition, the first eigenvalue λ_0 is always zero, and f_0 is a constant function. The spectrum is an isometric invariant because it only depends on the gradient and divergence, which are based only on the Riemannian structure of the manifold. After the normalization of the eigenvalues, the shape can be matched regardless of the scales. The inner product of the functions on M is defined by an integral over the manifold as

$$< f, g >= \int_M fg d\sigma. \tag{2.4}$$

The Laplace–Beltrami operator is Hermitian, so the eigenvectors corresponding to their different eigenvalues are orthogonal:

$$< f_i, f_j >= \int_M f_i f_j d\sigma = 0 \tag{2.5}$$

for $i \neq j$. According to the definition of the eigenvalue problem, if there exists a solution f_i, then αf_i is also a solution for every nonzero scalar α. Thus f_i is usually normalized as

$$< f_i, f_i >= \int_M f_i^2 d\sigma = 1 \text{ for } i = 0, 1, 2, \ldots. \tag{2.6}$$

A weak version of Eq. (2.3) can also be derived as

$$\int_M \nabla\varphi \cdot \nabla f d\sigma = \lambda \int_M \varphi f d\sigma \tag{2.7}$$

for any test function $\varphi \in C^1$, which is necessary but not sufficient for f to be a solution of Eq. (2.3). This weak version is proved by means of Green's formula on a closed manifold or an open manifold with Neumann boundary condition as

$$\begin{aligned} \Delta f &= -\lambda f, \\ \varphi \Delta f &= -\lambda \varphi f, \\ \int_M \varphi \Delta f d\sigma &= -\lambda \int_M \varphi f d\sigma, \\ -\int_M \nabla\varphi \cdot \nabla f d\sigma &= -\lambda \int_M \varphi f d\sigma. \end{aligned} \tag{2.8}$$

A given function f on the surface can be expanded as

$$f = c_1 f_1 + c_2 f_2 + c_3 f_3 + \cdots \tag{2.9}$$

with coefficients

$$c_i = < f, f_i > = \int_M f f_i d\sigma. \tag{2.10}$$

2.2 Finite element method computation

To solve the Laplace eigenvalue problem numerically, the manifold is discretized. Eq. (2.7) can be solved by the finite element method [116]. Assume that n sample points are given on the manifold M. Each function is defined on those n digital sample points. Differential operations are also defined on such points. If a set of n linear independent functions $\{\varphi_i\}$ is chosen, which means that

$$< \varphi_i, \varphi_j > = \int_M \varphi_i \varphi_j d\sigma = 0 \text{ if } i \neq j, \tag{2.11}$$

then any function on M, including the eigenfunction f, is a linear combination of $\{\varphi_i\}$ as

$$f = c_1\varphi_1 + c_2\varphi_2 + c_3\varphi_3 + \cdots + c_n\varphi_n. \tag{2.12}$$

The eigenfunction f is uniquely represented by the coefficient vector c and satisfies Eq. (2.7). If we consider each φ_i as a test function, then we get

$$\sum_{j=1}^{n} c_j \int_M \nabla\varphi_i \cdot \nabla\varphi_j d\sigma = \lambda \sum_{j=1}^{n} c_j \int_M \varphi_i \varphi_j d\sigma \text{ for } i = 1, 2, \ldots, n. \tag{2.13}$$

These n independent linear equations can be put into a linear system as

$$\begin{aligned} A \cdot c &= \lambda B \cdot c, \\ A_{ij} &= \int_M \nabla\varphi_i \cdot \nabla\varphi_j d\sigma, \\ B_{ij} &= \int_M \varphi_i \varphi_j d\sigma. \end{aligned} \tag{2.14}$$

Depending on different representations, $\nabla\varphi_i$ and the integral can be calculated on the finite elements. The weak version of the eigenvalue problem is converted into a generalized eigenvalue problem in matrix form. The family of solutions represents the eigenvalues and eigenfunctions.

2.3 Discrete Laplace–Beltrami operator

Another numerical approach is a discrete differential operator [94]. In our framework, 2D manifold data are approximated with discrete triangle meshes. A triangle mesh is defined with $M = (V, E, F)$, where $V = \{\mathbf{p}_i\}$ denotes the set of vertices, $E = \{\mathbf{e}_{ij}\}$ is the edge set, and $F = \{\mathbf{f}_{ijk}\}$ is the face set with

$1 \leq i, j, k \leq n = |V|$; $\mathbf{p}_i$ also denotes the position of the vertex in R^3, and the edge vector $\mathbf{e}_{ij}$ connects vertices $\mathbf{p}_i$ and $\mathbf{p}_j$ as $\mathbf{e}_{ij} = \mathbf{p}_j - \mathbf{p}_i$. The 1-ring neighbors of $\mathbf{p}_i$ are denoted as $N_1(i)$. All triangular faces assume counterclockwise orientation. Each triangular face represents a local manifold. We can define the property on each element, for example, vertex, edge, and face, which is a spatial average around such element. The properties on the vertices are considered as discrete samplings on the manifold. Discrete operators are also defined on each vertex. In this work the neighborhood of a vertex $\mathbf{p}_i$ is approximated by its Voronoi area. Fig. 2.1 demonstrates the Voronoi area of the vertex $\mathbf{p}_i$ within

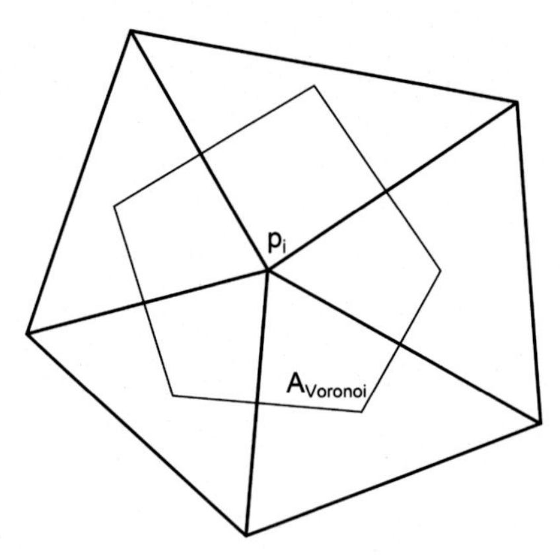

FIGURE 2.1 Voronoi area of the vertex $\mathbf{p}_i$ within its one-ring neighborhood. (Image taken from [42]. Included here by permission.)

its $N_1(i)$. A discrete Laplace–Beltrami operator K_i, also known as the mean curvature normal operator, is defined as the average value over the Voronoi area. Suppose we have the integral of the Laplace–Beltrami value. Then the Laplace–Beltrami operator at vertex $\mathbf{p}_i$ is represented as

$$K_i(g) = \frac{\int_{A_{\text{Voronoi}}} K(g)dA}{A_{\text{Voronoi}}}, \tag{2.15}$$

where $g(p) \in C^2$ is a scalar function on the triangle mesh M. The calculation of Voronoi areas on triangle meshes is trivial by definition. The geometry on each face is piecewise linear, so the Laplace–Beltrami operator turns into the Laplacian on the parameter space such that

$$K(g) = -\Delta g = -g_{uu} - g_{vv}, \tag{2.16}$$

where u and v are the parameters. According to Gauss' theorem, the area integral of the Laplacian can be calculated with a line integral on the boundary:

$$\int_{A_{\text{Voronoi}}} K(g)dA = -\int_{A_{\text{Voronoi}}} \Delta g dA = -\int_{\partial A_{\text{Voronoi}}} \nabla_{uv} g \cdot \mathbf{n} dl, \tag{2.17}$$

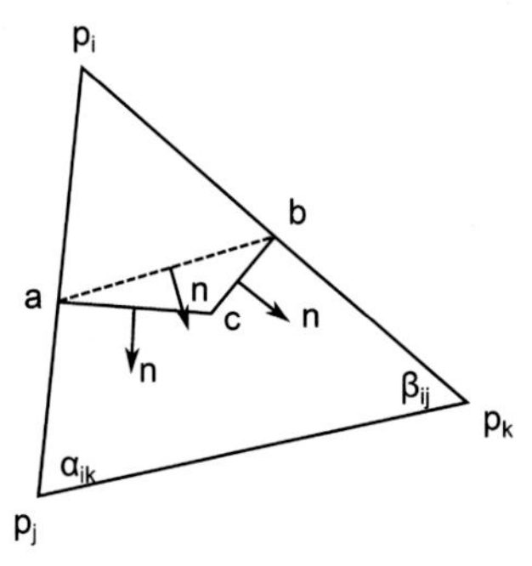

FIGURE 2.2 The boundary of a Voronoi area consists of a piecewise line segment in each face. (Image taken from [42]. Included here by permission.)

where $\mathbf{n}$ is the unit normal vector of the boundary. The boundary is a piecewise line segment in the one-ring neighbor faces, and the gradient is analytic and constant within each face. Fig. 2.2 demonstrates the piecewise linear boundary of the Voronoi area of $\mathbf{p}_i$ in the face $\mathbf{f}_{ijk}$. As we know, the gradient in each face is a constant vector, and the line segment integral on the segments ac and cb is equivalent to that on the dot straight line ab in the face. That is, in this face $\mathbf{f}$,

$$\int_{\partial A_{\mathrm{Voronoi}} \cap \mathbf{f}} \nabla_{uv} g \cdot \mathbf{n} dl = \nabla_{uv} g \cdot \mathbf{n}_{ab} \parallel \mathbf{e}_{ab} \parallel, \tag{2.18}$$

where a and b are the middle points of $\mathbf{e}_{ij}$ and $\mathbf{e}_{ik}$ by definition, so $\mathbf{e}_{ab}$ and $\mathbf{e}_{jk}$ are parallel, and $\parallel \mathbf{e}_{ab} \parallel = \frac{1}{2} \parallel \mathbf{e}_{jk} \parallel$. The normal direction of a line segment in a face is retrieved by rotating the directed edge 90° on the face plane. This rotation can be derived from the vector cross product. Denote by $\mathbf{N}$ the unit normal of the face $\mathbf{f}_{ijk}$:

$$\mathbf{N} = \frac{1}{2A_f} \mathbf{e}_{ki} \times \mathbf{e}_{ij} = \frac{1}{2A_f} \mathbf{e}_{ij} \times \mathbf{e}_{jk} = \frac{1}{2A_f} \mathbf{e}_{jk} \times \mathbf{e}_{ki}, \tag{2.19}$$

where A_f is the area of $\mathbf{f}_{ijk}$. Thus the rotation above is derived from the cross product $\mathbf{N}$ from the right, and Eq. (2.18) turns into

$$\int_{\partial A_{\mathrm{Voronoi}} \cap \mathbf{f}} \nabla_{uv} g \cdot \mathbf{n} dl = \frac{1}{2} \nabla_{uv} g \cdot (\mathbf{e}_{jk} \times \mathbf{N}). \tag{2.20}$$

The face is piecewise linear, and the gradient on it is constant and expressed as

$$\nabla_{uv} g = \frac{1}{2A_f} (g_i \mathbf{e}_{jk} + g_j \mathbf{e}_{ki} + g_k \mathbf{e}_{ij}) \times \mathbf{N}, \tag{2.21}$$

where g_i, g_j, and g_k are the values of a function g at those vertices. Combining equations (2.20) and (2.21), we have

$$\int_{\partial A_{\text{Voronoi}} \cap \mathbf{f}} \nabla_{uv} g \cdot \mathbf{n} dl = \frac{1}{4A_f}((g_j - g_i)\mathbf{e}_{ik} \cdot \mathbf{e}_{kj} + (g_k - g_i)\mathbf{e}_{ij} \cdot \mathbf{e}_{jk}), \quad (2.22)$$

or

$$\int_{\partial A_{\text{Voronoi}} \cap \mathbf{f}} \nabla_{uv} g \cdot \mathbf{n} dl = \frac{1}{2}(\cot \beta_{ij}(g_j - g_i) + \cot \alpha_{ik}(g_k - g_i)), \quad (2.23)$$

where α_{ik} and β_{ij} are the angles shown in Fig. 2.2. The sum of Eq. (2.23) of all the faces within the one-ring neighborhood leads to the discrete Laplace–Beltrami operator K_i on vertex $\mathbf{p}_i$

$$K_i(g) = \frac{1}{2A_i} \sum_{\mathbf{p}_j \in N_1(\mathbf{p}_i)} (\cot \alpha_{ij} + \cot \beta_{ij})(g_i - g_j), \quad (2.24)$$

where α_{ij} and β_{ij} are the two angles opposite to the edge in the two triangles sharing the edges i and j, and A_i is the Voronoi area of $\mathbf{p}_i$. For all vertices of a triangle mesh, the Laplace–Beltrami matrix can be constructed as

$$L_{ij} = \begin{cases} -\frac{\cot \alpha_{ij} + \cot \beta_{ij}}{2A_i} & \text{if } i, j \text{ are adjacent}, \\ \sum_k \frac{\cot \alpha_{ik} + \cot \beta_{ik}}{2A_i} & \text{if } i = j, \\ 0 & \text{otherwise}, \end{cases} \quad (2.25)$$

where α_{ij}, β_{ij}, and A_i are the same as in Eq. (2.24) for certain i and j. Then the spectrum problem (2.3) turns into the following eigenvalue problem:

$$L\mathbf{v} = \lambda \mathbf{v}, \quad (2.26)$$

where $\mathbf{v}$ is an n-dimensional vector. Each entry of $\mathbf{v}$ represents the function value at one of n vertices on the mesh. This equation can be represented as a generalized eigenvalue problem, which is much easier to solve numerically by constructing a sparse matrix W and a diagonal matrix S such that

$$W_{ij} = \begin{cases} -\frac{\cot \alpha_{ij} + \cot \beta_{ij}}{2} & \text{if } i, j \text{ are adjacent}, \\ \sum_k \frac{\cot \alpha_{ik} + \cot \beta_{ik}}{2} & \text{if } i = j, \\ 0 & \text{otherwise}, \end{cases}$$

and $S_{ii} = A_i$. Thus the Laplace matrix L is decomposed as $L = S^{-1}W$, and the generalized eigenvalue problem is presented as

$$W\mathbf{v} = \lambda S\mathbf{v}. \quad (2.27)$$

As defined before, W is symmetric, and S is symmetric positive definite. All the eigenvalues and eigenvectors are real, and the eigenvectors corresponding to different eigenvalues are orthogonal in terms of the S dot product

$$< \mathbf{u}, \mathbf{w} >_S = \mathbf{u}^T S \mathbf{w}, \tag{2.28}$$

where $\mathbf{u}$ and $\mathbf{w}$ are eigenvectors of Eq. (2.27). Eqs. (2.5), (2.9), and (2.10) can be reduced, respectively, to

$$< \mathbf{v}_i, \mathbf{v}_j >_S = 0, i \neq j, \tag{2.29}$$

$$\mathbf{v} = \sum_{i=1}^{n} \mathbf{v}_i c_i, \tag{2.30}$$

and

$$c_i = < \mathbf{v}, \mathbf{v}_i >_S . \tag{2.31}$$

Under this setting, the spectrum, $\{0, \lambda_1, \lambda_2, \lambda_3, \cdots, \lambda_{n-1}\}$ is the family of eigenvalues of the generalized eigenvalue problem defined previously.

Chapter 3

Spectral geometric features for shapes

Contents

In this chapter, we first introduce general geometric features of shapes and then a new method for extracting salient features from shapes. This method extracts salient geometric feature points in the Laplace–Beltrami spectral domain instead of usual spatial domains. Simultaneously, a spatial region is determined as a local support of each feature point, which corresponds to the "frequency" where the feature point is identified. The local shape descriptor of a feature point is the Laplace–Beltrami spectrum of the spatial region associated with the point, which is stable and distinctive. The method leads to the salient spectral geometric features invariant to spatial transformations such as translation, rotation, and scaling. The properties of the discrete Laplace–Beltrami operator make them invariant to isometric deformations and mesh triangulations as well. With the scale information transformed from the "frequency", the local supporting region always remains the same ratio to the original model no matter how it is scaled. This means that the spatial region is scale invariant as well. Therefore, both global and partial matchings can be achieved with these salient feature points as shown in Fig. 3.1.

Spectral Geometry of Shapes. https://doi.org/10.1016/B978-0-12-813842-7.00011-5

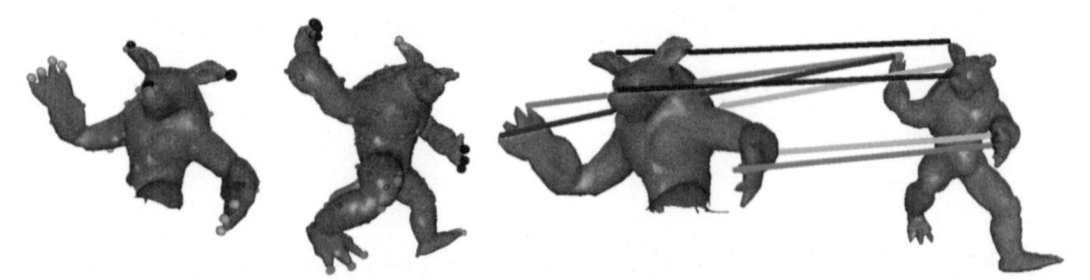

FIGURE 3.1 Matching with salient spectral geometric features. The highlighted feature points are extracted in the spectral domain generated directly from the triangle meshes. Notice that two meshes are different in position, orientation, scale, pose, number of vertices, and triangulation. Furthermore, the left shape is only a "part" of the whole model. To clearly see the matching result, only some of the matched points are displayed. (Image taken from [44]. ©2009 Springer Berlin Heidelberg. Included here by permission.)

3.1 General geometric features for shapes

There are a number of well-studied global shape representations such as moments [123,26,101], extended Gaussian images [41,132], shape distributions [105,49,103], and shape harmonics [60]. These representations show a great power in shape analysis. However, matching with these global representations usually requires the data to be aligned or normalized. It can be used to determine whether two shapes are similar or not, or how similar they are with those global representations, measured as a distance between the global shape representations. However, it is difficult to obtain local matching details directly, such as part-level similarity, correspondence, or registration. These global representations also perform poorly in a "part-to-whole" matching. A part or a subshape is always considered as a quite different shape from its original shape by these methods. In fact, a proper scale of the part to the whole shape itself is also regarded as a difficult task.

A part-to-whole matching is considered as a partial matching, which is more general and challenging than the global one. Partial matching decides whether a shape is a part of another one and where it should be located. It is often achieved with matching local features. Gal et al. [31] proposed a partial matching method based on salient local features extracted from 3D surfaces. The salient features are extracted locally with an area-growing algorithm following an empirical formula, and the descriptors are defined on the quadratic fitted surfaces based on the original meshes. Along a similar direction, there are also more rigorous scale space-based methods for extracting salient features from 3D surface shapes [76, 46,160]. A graph-based approach is another important solution to shape matching. For example, Reeb graph [39] and skeletal graph [137,50] represent a shape with a graph and turn the matching problem into a graph problem. A part-to-whole matching can then be handled with subgraph searching. However, the graph extracted from a shape is sensitive to its topology. The tiny change of topology may result in quite different graphs.

In this chapter, we discuss shape analysis based on the spectra of shapes. The shape spectrum is a new topic in computer vision and computer graphics in the recent years. As described in previous chapters, Reuter [116] defined the shape spectrum as the family of eigenvalues of the Laplace–Beltrami operator on a manifold and used it as a global shape descriptor. Lévy [79] pointed out that Laplace–Beltrami eigenfunctions are "tools" to understand the geometry of shapes and discussed the properties of the corresponding eigenfunctions of the Laplace–Beltrami operator. Rustamov [121] proposed a modified shape distribution based on eigenfunctions and eigenvalues. Karni and Gotsman [58] used the projections of geometry on the eigenfunctions for mesh compression. The Laplace–Beltrami spectrum is showing more and more power in shape analysis, and it has many desired invariant properties [116]. In this chapter, we mainly explain how to extract salient geometric features in the domain of shape spectrum.

3.2 Salient spectral features and extraction methods

Given a 3D triangular mesh of a shape, its Laplacian spectrum is computed as discussed in Chapter 2, and then the shape can be represented by a set of salient feature points with scale information in its spectrum. Local Laplacian spectra are calculated and assigned as local shape descriptors to the feature points together with their associated scales.

3.2.1 Salient feature point detection

In this section, we describe how to extract salient feature points based on the eigenvalues and eigenfunctions, which contain rich shape information. Fig. 3.2 illustrates some eigenfunctions on the armadillo model. The eigenfunctions contain very symmetric and meaningful information. Fig. 3.3 shows isometric properties of eigenfunctions. They are the 5th eigenfunctions on different meshes. The three meshes are generated from the same shape with different poses. The sampling rates are also different. In Fig. 3.3, (A) has 1000 vertices, (B) has 1500, whereas (C) has 3000. We can see that the eigenfunctions are independent to poses and triangulations.

(A)

(B)

(C)

(D)

FIGURE 3.2 The 2nd, 3rd, 4th, and 10th eigenfunctions on the shape. Red color indicates the larger value, whereas blue color denotes the smaller value. (Image taken from [44]. ©2009 Springer Berlin Heidelberg. Included here by permission.)

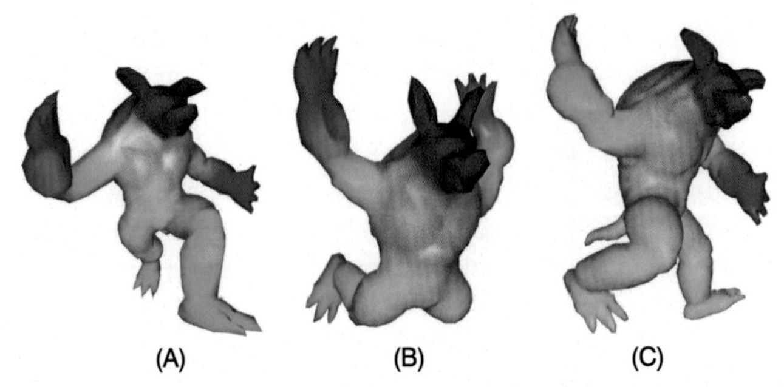

FIGURE 3.3 The 5th eigenfunctions on the model with different poses. The eigenfunctions are isometric invariant. The three shapes, from left to right, have 1000, 1500, and 3000 vertices, respectively. (Image taken from [44]. ©2009 Springer Berlin Heidelberg. Included here by permission.)

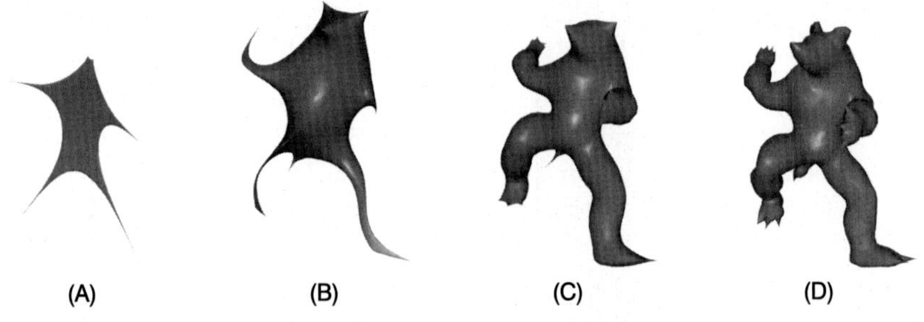

FIGURE 3.4 Geometric reconstructions with first 5 (A), 20 (B), 100 (C), and 400 (D) eigenfunctions. (Image taken from [44]. ©2009 Springer Berlin Heidelberg. Included here by permission.)

Consider the geometric reconstruction problem, and let $\mathbf{P}$ be the position matrix consisting of $\{\mathbf{x}, \mathbf{y}, \mathbf{z}\}$ coordinates of each vertex:

$$\mathbf{P} = [\mathbf{x}, \mathbf{y}, \mathbf{z}], \tag{3.1}$$

where $\mathbf{x} = [x_1, x_2, x_3, ..., x_n]^T$, $\mathbf{y} = [y_1, y_2, y_3, ..., y_n]^T$, and $\mathbf{z} = [z_1, z_2, z_3, ..., z_n]^T$ are coordinate vectors, and n is the number of vertices. Then the expansion with Eq. (2.30) and Eq. (2.31) can be rewritten in a matrix form as

$$\mathbf{P} = \mathbf{V}\mathbf{C}^T, \tag{3.2}$$

where $\mathbf{V}$ is the eigenvector matrix $\mathbf{V} = [\mathbf{v}_1, \mathbf{v}_2, ..., \mathbf{v}_n]$, and $\mathbf{C} = \mathbf{P}^T\mathbf{S}\mathbf{V}$. Let $\mathbf{A}_{1-p,1-q}$ denote the submatrix consisting of $1-p$ rows and $1-q$ columns of a matrix $\mathbf{A}$. Then the reconstruction of the first k eigenfunctions is represented as

$$\mathbf{P}(k) = \mathbf{V}_{1-n,1-k}\mathbf{C}^T_{1-3,1-k}. \tag{3.3}$$

The reconstructed mesh is represented by the coordinate $\mathbf{P}(k)$ with the same connections. Fig. 3.4 shows a reconstruction process. The eigenfunctions corresponding to smaller eigenvalues represent the lower-frequency information.

When an increasing number of eigenfunctions are included, more details of the mesh are presented. New salient features come out due to newly added eigenfunctions, which means that features must be contained in their eigenfunctions within the corresponding frequencies. Therefore we define the spectral geometry energy of a vertex i corresponding to the kth eigenvalue as

$$E(i,k) = \| \mathbf{V}(i,k) \times \mathbf{C}_{1-3,k} \|_2 . \tag{3.4}$$

The feature points are selected at maxima in E, which means that more spectral geometry energy is added locally in both spatial and spectral neighborhoods. If $E(i,k)$ is larger than those of its neighboring vertices within several neighbor frequencies, it will be finally chosen as a salient feature point with a scale factor $sf = 1/\sqrt{\lambda_k^2}$. Notice that one vertex may be selected several times with different scale factors corresponding to different eigenvalues. See Fig. 3.5 for an example.

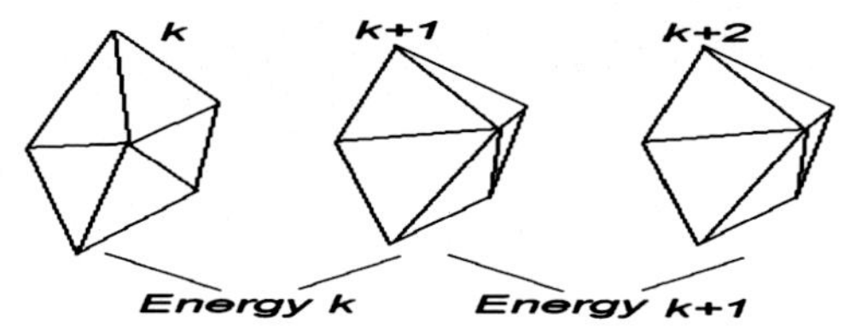

FIGURE 3.5 Spectral geometry energy between neighboring eigenfunction reconstructions. In this example the vertex in the middle of the one ring neighborhood receives the largest spectral geometry energy when the $(k+1)$th eigenfunction is added for reconstruction. Thus it is considered as a maximum at the "frequency" of λ_k. (Image taken from [44]. ©2009 Springer Berlin Heidelberg. Included here by permission.)

3.2.2 Shape descriptor construction

In previous sections, we have described how feature points are extracted from an original mesh with scale information factors. The next step is finding a local descriptor for each feature point, which is invariant to translation, rotation, scaling, and isometric deformation and also distinctive enough for similarity measure. The Laplace–Beltrami spectrum of the spatial local region defined by the identified scale (i.e., the local support of the corresponding salient feature point) can be employed. However, most of these regions are open boundary subsurfaces and properties of those eigenfunctions of the Laplace–Beltrami operator. Rustamov [121] mentioned that the Laplace matrix may encounter some problems with open boundary surface. To solve this problem, another surface patch is attached to the open boundary region patch. The attached patch has exactly the same shape as the original patch, but opposite normal at every point. Then an open boundary path turns into a water-tight surface and Eq. (2.27) can be applied on it without any problem.

The algorithm procedure is as follows: First, a spatial local patch is extracted by drawing geodesic circle with the feature point as the center and $r \times 1/\sqrt{\lambda_k^2}$

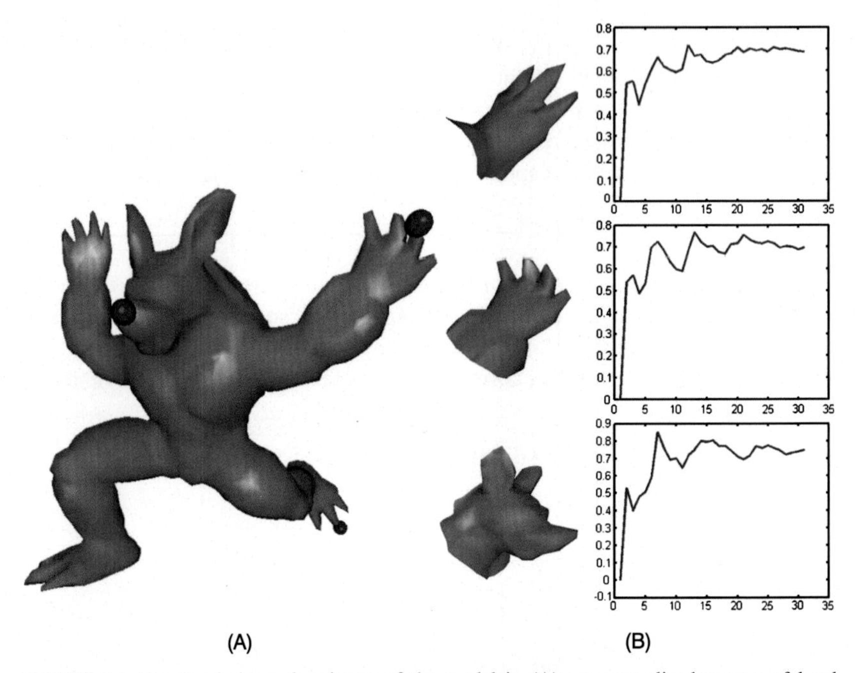

FIGURE 3.6 The local shape descriptors of the model in (A) are normalized spectra of local patches as shown in (B). (Image taken from [44]. ©2009 Springer Berlin Heidelberg. Included here by permission.)

as the radius, where r is a uniform, constant radius factor. Note that because of the scale factor $1/\sqrt{\lambda_k^2}$, the shape of the local patch will remain the same despite the scaling of the mesh. Then, Eq. (2.27) is applied on the patch to obtain the spectrum of the patch. Finally, for similarity comparison, the spectrum is normalized by a fitting function $f(x) = \frac{4\pi}{A}x, x = 1, 2, ..., n$, where A is the area of the local patch [116]. Fig. 3.6 illustrates how to construct a descriptor for a feature point. The histograms show the spectral values over the eigenvectors in the local patches. The matching process can be performed by comparing the Euclidean distance between two descriptors.

3.3 Spectral feature-based shape correspondence

Given a 3D surface, we can now represent it with a set of salient spectral geometric features. In this section, we discuss a method of building a correspondence between two models with those features. The correspondence problem can be described as follows: if there are two sets of features $\{p_i\}$ and $\{p'_{i'}\}$, try to find a mapping function ϕ such that $\phi(i) = i'$. The similarity between these two models relies on the mapping function ϕ. We denote the similarity as

$$Sim(\phi) = Sim_s(\phi) + Sim_p(\phi), \tag{3.5}$$

where $Sim_s(\phi)$ is the similarity calculated based on a single-feature-to-single-feature mapping, and $Sim_p(\phi)$ is based on a cluster-to-cluster mapping. The single-to-singe feature similarity can be computed based on the Euclidean distances between the spectrum descriptors. Let $f(i)$ denote the feature vector of the feature i. Then $Sim_s(\phi)$ is defined as

$$Sim_s(\phi) = \omega_s \sum_{\phi(i)=i'} C(i,i'),$$
$$C(i,i') = \exp\left(\frac{\| f(i) - f(i') \|^2}{2\sigma_s^2}\right). \tag{3.6}$$

The cluster-to-cluster feature similarity can be computed based on the relative geodesic distances and scale factors. Let $g(i, j)$ denote the absolute geodesic distance between i and j based on the spatial coordinate, and let $sf(i)$ denote the scale factor of i. Then $Sim_p(\phi)$ is defined as

$$Sim_p(\phi) = \omega_p \sum_{\phi(i)=i',\phi(j)=j'} H(i,j,i',j'),$$
$$H(i,j,i',j') = \exp\left(\frac{(d_{pg}(i,j,i'j') + \beta d_{ps}(i,j,i'j'))^2}{2\sigma_p^2}\right), \tag{3.7}$$

where d_{pg} is the distance between relative geodesic distances

$$d_{pg}(i,j,i'j') = \left|\frac{g(i,j)}{sf(i)} - \frac{g(i',j')}{sf(i')}\right|, \tag{3.8}$$

d_{ps} is the distance between scale ratios

$$d_{ps}(i,j,i'j') = \left|\log\left(\frac{sf(j)}{sf(i)}\right) - \log\left(\frac{sf(j')}{sf(i')}\right)\right|, \tag{3.9}$$

and ω_s, σ_s, ω_p, σ_p, and β are weight scalars. The goal of the correspondence algorithm is finding a mapping function ϕ_c that maximizes the similarity $Sim(\phi)$. If we define binary indicators variable $x(i,i')$ as

$$x(i,i') = \begin{cases} 1 & \text{if } \phi(i) = i' \text{ exists,} \\ 0 & \text{otherwise,} \end{cases} \tag{3.10}$$

then Eq. (3.5) can be represented with an Integer Quadratic Programming (IQP) problem as

$$Sim(\mathbf{x}) = \omega_p \sum_{i,i',j,j'} H(i,j,i',j')x(i,i')x(j,j') + \omega_s \sum_{i,i'} C(i,i')x(i,i'). \tag{3.11}$$

A constraint on a one-to-one mapping, which means that one feature in a model cannot be assigned more than one correspondence in the other model, needs to be enforced. Consequently, we have $\sum_i x(i, i') \leq 1$ and $\sum_{i'} x(i, i') \leq 1$. These linear constraints can be encoded in one row of **A** and an entry of **b**. Therefore our IQP problem can be formalized in the following matrix form:

$$\max Sim(\mathbf{x}) = \mathbf{x}'\mathbf{H}\mathbf{x} + \mathbf{C}\mathbf{x} \text{ subject to } \mathbf{A}\mathbf{x} \leq \mathbf{b}. \tag{3.12}$$

The IQP solver proposed by Bemporad et al. [6] can be used to solve this optimization problem.

3.4 Experiments and applications on shape matching

Now we use some experiments and applications to help us getting familiar with the discussed methods and algorithms.

Different from traditional image processing, 3D shape analysis replies more on the geometry, usually represented by irregularly sampled triangle meshes. 3D shapes also have more degrees of freedom and deformations. The shape spectrum is invariant to translation, rotation, and scaling as well as to triangulations and near-isometric deformation. It does well in understanding the shapes globally. The spectrum itself is also a good shape descriptor for shape matching and retrieval. However, there are still some limitations for this approach: (1) Near isometry is a strict assumption and cannot be guaranteed in many real cases, for example, in large shape database retrieval; and (2) it is a global feature and helps less in local or partial matching. In Fig. 3.7 the spectra on the near-isometric camel shapes are quite invariant, although they have different triangulations and poses. On the other hand, a similar four-limb horse has a spectrum different from that of a camel. Our goal is finding local-based salient features in the global spectrum. They carry the same invariance and overcome the limitations of the original spectrum.

FIGURE 3.7 The columns represent the 2nd, 3rd, 4th, 5th, and 6th eigenfunctions.

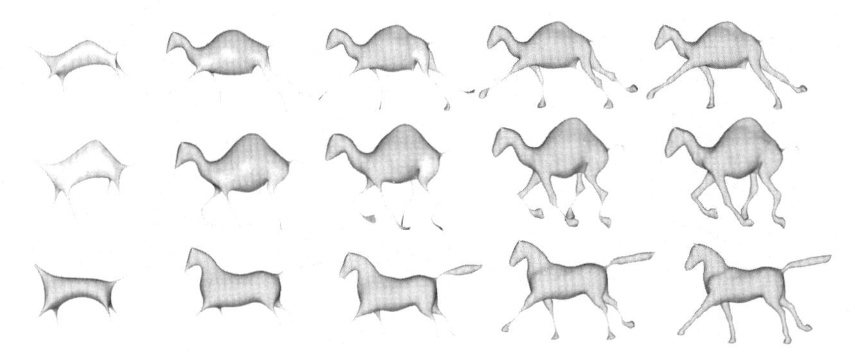

FIGURE 3.8 The columns represent the reconstructions from the first 10, 50, 100, 200, 500 eigenenergies.

We present the spectrum energy from a spectrum reconstruction process. The well-known Fourier transform is a particular case of the Laplace–Beltrami operator on 2D plane domain. If we define the spatial geometry as signals, eigenvalues as frequencies, and eigenvectors and basis, then the traditional signal processing also applies to the Laplace–Beltrami spectrum space. The previous section already demonstrates a typical spectrum reconstruction process. In this one, we give more examples to show the similarities between the near-isometric and nonisometric shapes during spectrum process. Fig. 3.8 shows, among different poses, triangulations, and even nonisometric shapes, the reconstructions sharing some similarities. The inspection introduces the presented salient spectral geometric features.

Salient spectral geometric features. In previous sections, we have discussed a method for extracting salient spectral geometric features from a surface shape. Fig. 3.9 shows some examples of salient feature points. The vertices colored with red and green colors are feature points extracted in the spectral domain with our method. Each mesh in Fig. 3.9 has 1000 to 1500 vertices. To find extrema in the spectral domain, each vertex is compared with its one-ring neighbors within three frequencies, that is, the current, previous, and next ones. The extrema are extracted in the first 100 eigenfunctions. A redder color means that the feature is found in a lower "frequency", which has large supporting region, whereas a greener color corresponds to a higher "frequency". Only the lowest "frequency" is visualized for a vertex with multiple "frequencies". The highlighted patches illustrate the local supports for some identified feature points. The experiments show that the salient features are very stable and invariant to Euclidean transforms and isometric deformations.

Shape correspondence. A nature application with the salient features is partial matching. Because each feature point is found with both position and scale factors, there is a local region to support the feature point. The ratio of the local support region to the entire surface is independent of the scale of the original surface. In our experiments, the radius constant r is set to be 1.7, and a

FIGURE 3.9 Salient feature points extracted in the spectral domain. A redder color means that the feature is found in a lower "frequency", which has a large supporting region, whereas a greener color corresponds to a higher "frequency". The highlighted patches illustrate the local supports for some of the feature points. (Image taken from [44]. ©2009 Springer Berlin Heidelberg. Included here by permission.)

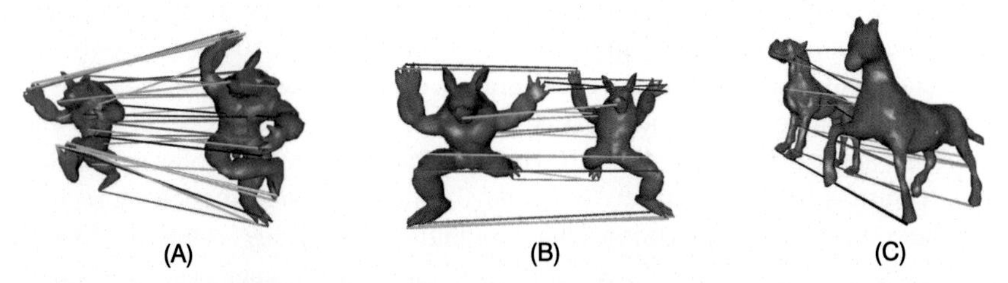

FIGURE 3.10 Examples of matching. (A) and (B) demonstrate a correspondence between shapes form the same model with different poses. (C) shows a correspondence between similar shapes. (Image taken from [44]. ©2009 Springer Berlin Heidelberg. Included here by permission.)

geodesic circle patch is approximated with the graph shortest edge path on the mesh. Then the spectrum of the local patch is calculated, and the descriptor consists of the eigenvalues of λ_1 to λ_{21} with normalization. Figs. 3.1 and 3.10 show some examples of global matching and partial matching. The two meshes

FIGURE 3.11 Examples of shape retrieval with salient spectral geometric features. The 3D shapes at the most left column are input queries, and those on the right are the first five retrieved results from the database. (Image taken from [44]. ©2009 Springer Berlin Heidelberg. Included here by permission.)

in the matching pairs are different in positions, orientations, scales, poses, and triangulations. The experiments demonstrated that the salient features are very powerful in matching of similar shapes. In Fig. 3.10, (A) and (B) are the shapes of the same armadillo model with different poses. We can see that even poses are quite different from each other but the correspondences are stable, even at very detailed levels. Note that the mirrored matches may happen in the algorithm. Not only the different poses from the same model, but also similar shapes can

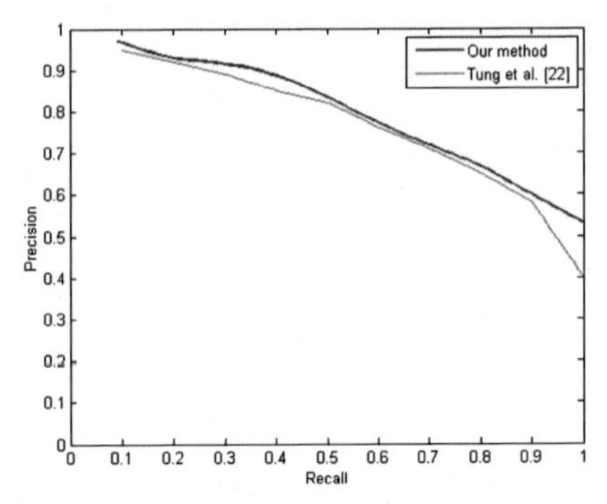

FIGURE 3.12 The overall averaged precision/recall graph of our method on the SHREC datasets and its comparison to the method by Tung et al. [140]. (Image taken from [44]. ©2009 Springer Berlin Heidelberg. Included here by permission.)

have correct correspondences. (C) illustrates that a dog shape corresponds to a horse shape with their similar features, such as heads, knees, necks, and feet.

Shape retrieval. Another application is 3D shape searching and retrieval in large databases. Not only globally the shape can also be searched by parts, which leads to a more powerful partial matching. We use SHREC 3D shape database.[1] We extract 10 shapes from each category. Since our framework extracts stable and distinctive salient spectral geometric features, the retrieval task is very straightforward by comparing the matching score of the IQP throughout the database and then selecting the best ones as the query outputs. Fig. 3.11 shows some results demonstrating the stability and accuracy of our retrieval. For the dataset that we use, our method outperforms the reported best result in the latest SHREC contest [140]. Fig. 3.12 shows the precision/recall graph.

As we mentioned in the previous sections, another desirable property of the salient spectral geometric features is their powerful partial representation. Fig. 3.13 shows how our method performs in retrieving shapes if only a part is given.

3.5 Summary

We have introduced a novel 3D shape representation with a set of salient feature points in the Laplace–Beltrami spectrum. The spectrum is defined as the family of eigenvalues of the Laplace–Beltrami operator on a manifold. The eigenvalues and eigenfunctions are invariant to translation, rotation, and scaling. They are also invariant to isometric transformations. We have discussed how to calculate the spectrum directly on triangle meshes. The results showed that the spectrum relies only on the geometry of the manifold. It is very stable under Euclidean and isometric transformations and is independent of different triangulations. The spectrum energy domain is obtained by projecting the geometry

1. The database can be obtained at the link: http://www.shrec.net/.

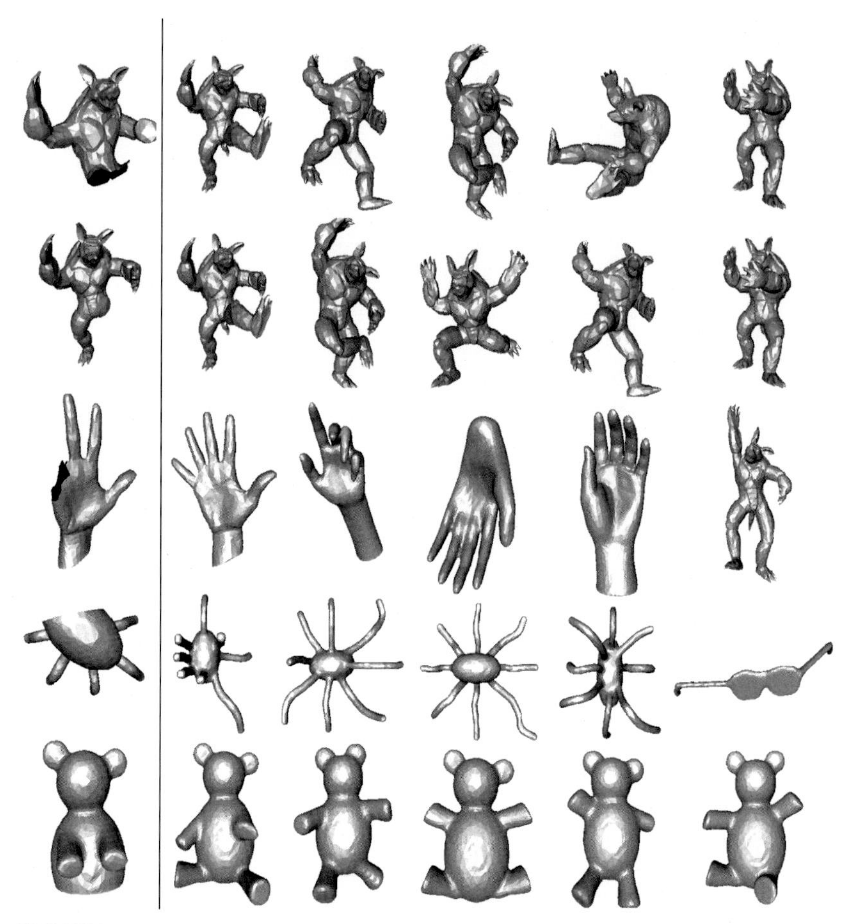

FIGURE 3.13 Examples of partial shape retrieval with salient spectral geometric features. The shapes at the most left column are input queries, and those on the right are the first five retrieved results from the database. (Image taken from [44]. ©2009 Springer Berlin Heidelberg. Included here by permission.)

onto the eigenfunctions. The salient features are the energy maxima in the geometry energy domain and share the nice properties of the Laplace–Beltrami spectrum. The maxima provide not only where the features are on the manifold but also the frequency where the features lie in. A scale of a local region can be determined with the frequency to support a feature point. With the IQP algorithm, correspondences can be built among variant shapes in very detailed levels. Salient spectral feature point representation is an ideal method for fundamental shape matching. The experiments have showed its great power in shape retrieval and searching. Besides global matching, partial matching can also be supported in this framework.

Chapter 4

Near-isometric motion analysis using spectral geometry

Contents

The previous chapter solves the problem of measuring the similarities among different static shapes. Poses casted by the same original object are considered as the same shape in terms of spectral features. However, there is still isometric deformations that carry rich geometric information. In this chapter, we present a novel method to analyze a set of poses of 3D models that are represented with triangle meshes and unregistered. Different shapes of poses are transformed from the 3D spatial domain to a geometry spectral domain defined by the Laplace–Beltrami operator. During this space-spectrum transform, all near-isometric deformations, mesh triangulations, and Euclidean transformations are filtered away. The different spatial poses from a 3D model are represented with near-isometric deformations, and therefore they have similar behaviors in the spectral domain. Semantic parts of that model are then determined based on the computed geometric properties of all the mapped vertices in the geometry spectral domain. A semantic skeleton can be automatically built with joints detected as well. The method turns a rather difficult spatial problem into a spectral problem that is much easier to solve.

4.1 Near-isometric shape deformation and motion

Shape animation and deformation often relies on shape interpolation. Given two or more key frames of a shape, the intermediate deformations are interpolated

Spectral Geometry of Shapes. https://doi.org/10.1016/B978-0-12-813842-7.00012-7

FIGURE 4.1 The procedure of our pose analysis method. Given several unregistered poses of a model that have different triangulations (shown in the left pane), an reembedding from the spatial domain to a geometry spectral domain is built as shown in the middle. The poses are analyzed in the geometry spectral domain. The geometric behavior of each point on the pose surface is classified. Then semantic parts on any poses from the same model can be determined. The colder color in the middle figure indicates a rigid part on the surface, whereas the warmer color denotes an articulated part. With the graph and skeleton driven algorithms, the static 3D surface turns into a semantically articulated model, which can cast animation. (Image taken from [45]. ©2013 Springer Berlin Heidelberg. Included here by permission.)

or blended. Extrapolation can also be applied, which decides what the shape is going to be following the deforming direction. The interpolation is applied on locations of the corresponding vertices or faces. Certain constrains are considered to make the interpolation as natural as possible while avoiding some artifacts such as local shrinking or collapse. Kilian et al. [63] treated each pose of shapes as a point in a shape space. A best interpolation is a geodesic path between two poses, which can preserve the original length on the surfaces as much as possible. James and Twigg [52] and Chu et al. [21] employed mean shift clustering to learn the near-rigid parts of surface from a sequence of poses to guide the interpolation. This kind of methods usually requires one-to-one vertex-face correspondence, either pregiven or obtained by other registration algorithms. The correspondence requirement limits the capabilities of these methods since registration itself is another challenging problem. The shape interpolation focuses on global smoothness and influence to provide smooth and fluent shape sequences. It is based on signal interpolations with constrains.

Skeleton-driven mesh deformation is another popular kind of shape approaches. A shape is deformed under the control of an articulated structure, which is more natural to human understanding [23]. It can provide local control and free deformation. Yan et al. [152] employed simplex transformations to make the skeletons drive the surfaces instead of vertices. Weber et al. [146] used geometric information to build the skeleton to preserve local details. This kind of approaches usually requires the skeletons to be manually designed to reach a better result. He et al. [38] introduced a harmonic function on the surface to build a Reeb graph [15,16,22,108,109,131] as the skeleton. They reduced the manual operation to picking only one or a few reference points on the surface.

All works aforementioned are from a perspective of spatial analysis. They have to overcome many Euclidean factors such as translation, rotation, and scaling before they analyze the pure geometry properties. Recent research shows that 3D surfaces can also have spectral properties, to which the Euclidean factors are not significant. Karni and Gotsman [57] defined the mesh Laplacian on polygon meshes based on the adjacent matrix. The eigenvectors of the Laplacian matrix form an orthogonal basis on the mesh surface. The eigenvalues denote different frequencies. Then signal processing algorithms can be applied to the mesh surface, such as filtering, denoising, and mesh compression. This mesh Laplacian relies on the triangulations of meshes. Reuter et al. [116] introduced the Laplace–Beltrami operator on Riemannian manifolds represented with surfaces in the 3D Euclidean space. The operator is invariant to Euclidean transformations and isometric deformations. The eigenvalues can be used as shape descriptors that are not only invariant but also distinctive. The eigenvalues also contain much information such as the area of the surface, topology, and boundary length. Lévy [79] focused more on the eigenfunctions of the Laplacian equation. The eigenfunctions form an orthogonal basis for the functions defined on the Riemannian manifold and can "understand the geometry". A lot of applications can be achieved, such as signal processing on surfaces, geometry processing, pose transfer, and parameterization. Rustamov [121] defined a Global Shape Descriptor (GPS) embedding based both on eigenvalues and eigenfunctions and gave a G2 distribution based on the GPS, which can be used as a global shape descriptor stable to topology changes. Hu and Hua [44] analyzed shapes with salient features extracted from the shape spectra.

Our work starts from the perspective of spectral geometry. Therefore it may extract the pure geometric information behind variant Euclidean factors. The discrete setting makes the Laplace–Beltrami operator applicable on a triangle mesh directly. This saves preprocessing and handles more types of data.

4.2 Shape spectrum on triangle meshes

Considering equation (2.3), the solution $\{0, \lambda_1, \lambda_2, \lambda_3, \ldots, \lambda_{n-1}\}$, is the family of eigenvalues of the generalized eigenvalue problem defined before. The eigenvectors $\mathbf{v}_0, \mathbf{v}_1, \mathbf{v}_2, \ldots, \mathbf{v}_{n-1}$ represent the eigenfunctions on the mesh. They define the spectrum of a shape. As it can be seen, the number of eigenvalues and eigenfunctions is reduced from infinity to n, because a triangle mesh is a finite discrete sampling of a continuous surface. It is similar to the discrete and continuous Fourier transforms. In practice, infinitely many eigenvalues and eigenfunctions are not necessary. Only a few first eigenvalues and eigenfunctions are employed to build the geometry spectral domain.

Fig. 4.2 illuminates the 3rd, 5th, and 10th eigenfunctions on different poses. Note that the Laplace–Beltrami operator is defined on a continuous manifold, so the triangle meshes are required to be manifolds. They could be either closed manifolds or those with open boundaries and the same topology. The color turns from cold to warm while the function value grows from a small one to a big one. The eigenfunctions always change along the surface geometry. The three poses are quite different from the spatial view, but the eigenfunctions stay stable on the surfaces. The eigenfunctions rely only on the surface geometry. The shapes are

FIGURE 4.2 The 3rd, 5th, and 10th eigenvectors of discrete Laplace matrices on three different poses. Each column demonstrates a pose, whereas each row shows the 3rd, 5th, and 10th eigenvectors from the top to the bottom. The color from blue to green and then to red demonstrates the value changes from small to large. Each eigenvector shows some meaning of the surface. Within a pose, a higher-order eigenvector shows a higher frequency. Note that the pose surfaces in the first column have about 2,000 vertices; those in the second column have about 10,000 vertices; and those in the last column have about 20,000 vertices. The eigenvectors are not only meaningful but also stable to poses and triangulations. (Image taken from [42]. Included here by permission.)

not only different from each other with poses, but also the triangulations. The pose in the first column has about 2,000 vertices; the one in the second column has about 10,000; the one in the last column has about 20,000. As discussed before, the eigenfunction is also invariant to triangulations. These properties guarantee that the geometry spectrum embedding is invariant to pose deformations and mesh triangulations. In other words, vertices from different poses but

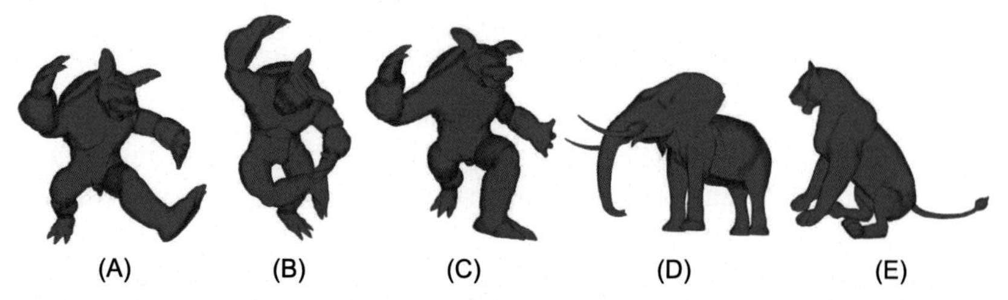

FIGURE 4.3 Five different shapes in the database. The first three shapes are different poses from the same armadillo model. According to Table 4.1, the three armadillo poses have similar eigenvalues, whereas the eigenvalues of the elephant and the lion are quit different. (Image taken from [45]. ©2013 Springer Berlin Heidelberg. Included here by permission.)

TABLE 4.1 Normalized eigenvalues of different shapes in Fig. 4.3. (Table taken from [45]. ©2013 Springer Berlin Heidelberg. Included here by permission.)

Shape	λ_0	λ_1	λ_2	λ_3	λ_4	λ_5	λ_6	λ_7
Armadillo (A)	0	1	1.23	1.64	2.90	4.37	6.32	8.83
Armadillo (B)	0	1	1.36	1.81	3.20	4.52	6.48	8.51
Armadillo (C)	0	1	1.25	1.33	2.28	4.83	6.76	8.68
Elephant (D)	0	1	2.44	3.07	3.51	3.98	4.24	4.70
Lion (E)	0	1	1.51	2.57	2.66	2.71	4.69	7.92

at the same position in terms of surface geometry will be embedded together in the geometry spectral domain, no matter how the poses are deformed or how different the samplings and triangulations are.

The spectrum can describe the intrinsic geometry within the original shape. Theoretically, the shape spectrum is invariant to isometric deformations. However, problems arise when dealing with real data. Different poses casted by an object are usually near isometric to each other. The deformations near the joints break the isometric constraint. The computations also bring numerical errors. Dey et al. [25] studied the spectral stabilities under near-isometric deformation. Their results show the spectra achieved with the cotangent scheme, including the discrete operator in our method, are stable in terms of eigenvalues. Our method produces similar results. Fig. 4.3 lists five shapes represented with triangle meshes, whereas Table 4.1 lists a few their first eigenvalues. The eigenvalues are normalized with the first nonzero eigenvalue to filter away the global scaling according to [116]. Because the first three poses are casted by the same armadillo model, they are considered to be near isometric to each other. This fact is demonstrated by the similar eigenvalues. When the models are different, the eigenvalues are dramatically different too, as shown in Table 4.1. The eigenvalues have enough power to distinguish models and shapes globally. In addition, there are some other potential problems of eigenvectors/eigenfunctions as discussed further, which may affect our algorithm. Rueter et al. also discussed these problems in [118].

- Sign flips occurs. If $\mathbf{v}$ is an eigenvector, then so is $-\mathbf{v}$ according to the definition. Rueter [118] admitted that sign flips cannot be detected intrinsically on an almost perfect intrinsic symmetric shape. We employ the absolute value to avoid the sign flip problem.
- Eigenvectors switch. The neighbor eigenvalues may switch due to the perturbations of the deformations and numerical computations. So are the corresponding eigenvectors. It happens nearly on every mesh. Rueter [118] gave a solution to reorder the eigenvectors based on the Morse–Smale graph. We use the same scheme. Without further mentioning, all the eigenvectors in the rest of the chapter refer to the reordered ones.
- Higher-dimensional eigenspaces can theoretically occur. However, this rarely happens in practical data. We have not found any example in our results so far.
- Duplicated eigenvalues may exist. A highly symmetric shape, for example, a sphere or cube, has duplicated eigenvalues. The linear combinations of the corresponding eigenvectors are also eigenvectors. Nevertheless, practically used animation models have no such high symmetry, that is, duplicated eigenvalues rarely happen practically in our application.
- Low-frequency eigenvectors are stable under near-isometric deformation. Rueter [118] had a detailed discussion on the stabilities of the shape spectrum with respect to near-isometric deformations and noises and used direct spectral embedding for the semantic shape segmentation. Our experiments also show that the low-frequency eigenvectors are quite stable. The third row of Fig. 4.2 demonstrates the stability of the 10th eigenvectors for different poses. Although the spectra are stable globally, the local values of an eigenvector may shift. This usually happens when there is a twisting deformation. The shifting affects the registration accuracy under this single frequency. However, with multifrequency embedding and multiple shape data, the accuracy are corrected by other values that are stable.

This discussion shows that the near-isometric shapes will have a similar behavior in the spectral spaces.

4.3 Spectral embedding of shape variation

The Laplace–Beltrami operator defines a family of eigenvalues and a family of eigenfunctions. The eigenvalues can be used as shape descriptors, which are stable and distinctive. It also contains "frequency" information. The smaller eigenvalues denote lower frequencies. The eigenvectors form an orthogonal basis on the manifold. Any functions can be projected to the basis and reconstructed by a linear combination of these eigenfunctions. All these are global analysis in the spectral domain. However, the main goal here is to study the local behaviors of the surfaces. It is obvious that each eigenfunction f_k is assigned a real value at every surface point p as $f_k(p)$. With respect to each point, there exists a mapping from a point on the surface in 3D spatial space to an infinite geometry spectrum space,

$$G_S(p) = \left(\frac{f_1(p)}{\sqrt{\lambda_1}}, \frac{f_2(p)}{\sqrt{\lambda_2}}, \frac{f_3(p)}{\sqrt{\lambda_3}}, \dots \right), \tag{4.1}$$

where p is a point on a surface S, and f_k is the kth eigenfunction corresponding to the kth eigenvalue λ_k of S. Each eigenfunction is normalized by

$$< f_i, f_i >= 1, i = 1, 2, 3, \dots, \tag{4.2}$$

on the surface S. As we work on the poses casted by the same object, the scales of each surface can be normalized. Thus the scales of the values in eigenfunctions represent the geometries of the shapes. We summarize some major advantages of this geometry spectral domain embedding as follows:

- If the surface in 3D space has no self-intersection, then the embedding has no self-intersection in the infinite domain either, which means that $G_S(p_i) = G_S(p_j)$ if and only if $p_i = p_j$ on S.
- The embedding is based only on eigenfunctions on the manifold. It only relies on the manifold metric and is invariant to the Euclidean embedding in the 3D space of S. The embedding mapping filters away the Euclidean transformations and near-isometric deformations.
- The embedding is invariant to different triangulations because of the implementations of the discrete Laplacian.

With this embedding, our method does not require any preprocessing such as normalization, remeshing, or registration. All these spatial factors do not matter in the geometry spectral domain. With the same surface, there is only one basis set for the embedding. Among different poses, the basis keeps stable. Fig. 4.4 demonstrates different poses registered in the spectral domain neutrally. The 3rd, 4th, and 5th eigenfunctions are picked for form a 3D subspace in the spectral domain. The Euclidean differences, including triangulations, transforms, and isometric deformations, are filtered away. The shapes are colored with the 3rd eigenfunction for rendering. The eigenfunctions of the manifold satisfies Eq. (2.3). If a certain function f_k is a normalized eigenfunction corresponding to some eigenvalue λ_k, according to this equation, $-f_k$ is also a normalized eigenfunction. The experiments also show that eigenfunctions from different poses can flip with sign corresponding to the same eigenvalue. To overcome this flipping problem, the mapping is restricted as an absolute one as

$$AG_S(p) = \left(\| \frac{f_1(p)}{\sqrt{\lambda_1}} \|, \| \frac{f_2(p)}{\sqrt{\lambda_2}} \|, \| \frac{f_3(p)}{\sqrt{\lambda_3}} \|, \dots \right). \tag{4.3}$$

The absolute mapping will break the first property about self-intersection, and the symmetric points on the surface will be mapped together. In our framework, it is natural to assume that parts have similar physical behaviors when they are

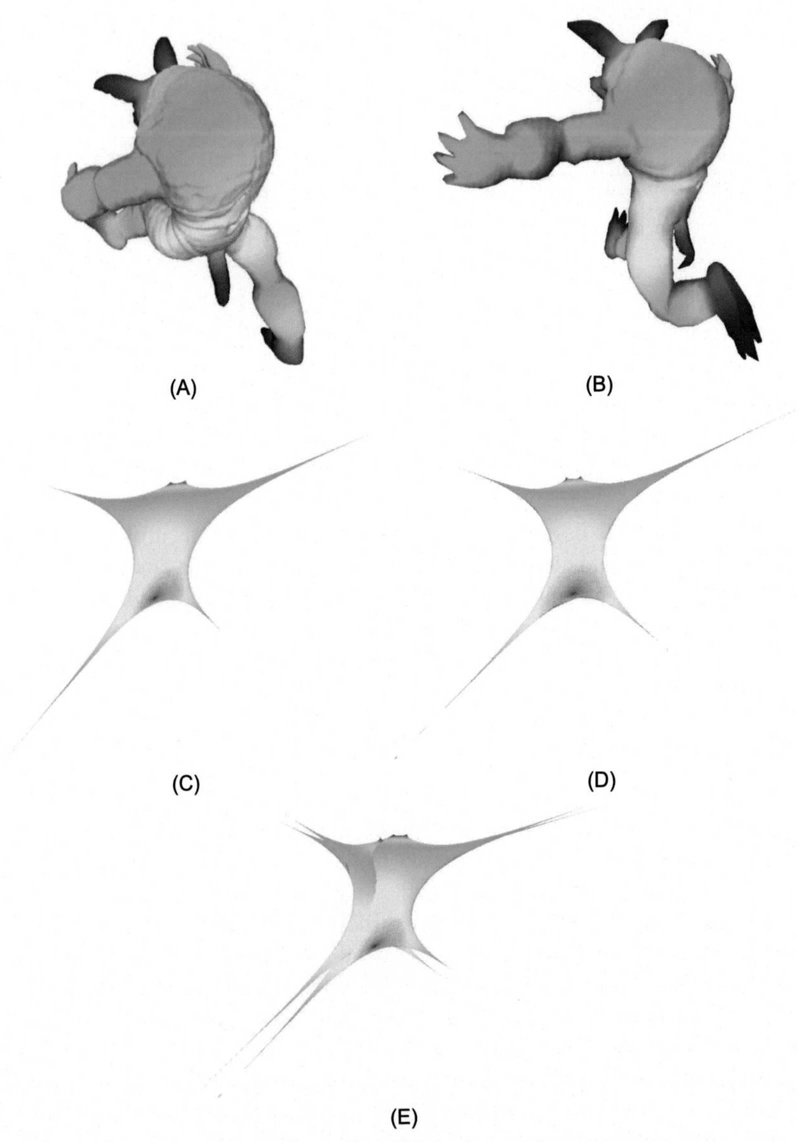

FIGURE 4.4 Transfer spatial manifolds to spectral domain. In the high-dimensional space the near-isometric deformation will be filtered out and registered uniformly. For rendering purpose, only the 3–5 eigenfunctions are chosen as a 3D subspace of the spectral domain. Despite the spatial deformations and different triangulations, the shapes are registered in the spectral domain neutrally. The color illustrates the 3rd eigenfunction distribution on shapes. (Image taken from [42]. Included here by permission.)

symmetric on the surface. Thus the absolute mapping does not affect the accuracy of mapping in terms of symmetry. For example, the left and right elbows have the symmetric geometry properties, and the registration across each is acceptable, as we do not require the dense mapping and registration.

4.4 Semantic shape analysis

4.4.1 Semantic point classification

The geometry spectrum embedding transforms each point on the surface from the Euclidean space to an infinite geometry spectrum space. Suppose there exists a spatial surface that is near-isometrically deformed along time, denoted by $S(t)$. The points with same positions relative to the surfaces S at different time will map into the same coordinates in the spectrum space, despite different locations, orientations, and poses of the original surfaces. Although the point is fixed in the spectral domain, it can carry varying geometric properties on S. That is to say, in the spectral domain, the properties at a point vary while the pose changes. For each point p in the spectral domain, we can define a property function $f_p(t)$ that depends only on time t. Imagine that if the properties are chosen to be invariant to Euclidean transformations but only sensitive to pose changes, what can be observed in the spectral domain is that properties vary on certain regions along with pose changes whereas they do not on the other regions. The former situation indicates articulations, whereas the latter one indicates rigid parts of the original shape. There are some well-studied features on surfaces, such as curvatures, normals, geodesic fans, and so on. In our framework, mean curvatures are a straightforward choice, as the Laplacian operator is also the mean curvature normal operator on the surface. When the shape deformation $S(t)$ is given, the pose behaviors of all the points can be classified into articulate or rigid.

The data in our framework are not a continuous surface changing with time but N frames of meshes, where N can range from 2 to 10 or even more. The property functions are reduced to a discrete set. A triangle mesh is a discrete sampling of a surface, and therefore an exact correspondence of a vertex may not exist on another near-isometric mesh. Thus the property set on a vertex is built based on an approximation. Suppose f_p is a feature set that is going to be built at a vertex p on the surface S. First, the embedding of p is calculated, and the mean curvature of p is put into f_p as an element. Then for each following frame of meshes S_i, a point p_i is found as a nearest one to p in the geometry spectral domain based on the Euclidean distance, and the mean curvature of p_i is put into f_p as another element. Therefore the element in f_p can classify the geometry behaviors of p through different poses. Fig. 4.5 illustrates the maxima, minima, and range distributions on the surface among different poses.

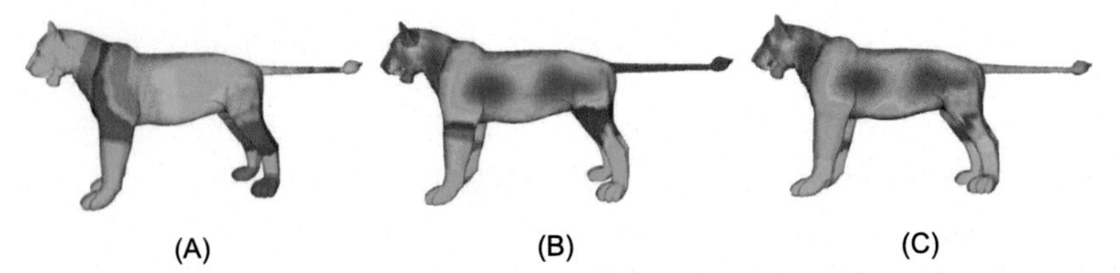

FIGURE 4.5 Mean curvatures values in spectral pose analysis. (A) maximum mean curvature distribution on each vertex during pose transformations; (B) minimum mean curvature distributions; (C) mean curvature range distribution. The values are histogram equalized for visualization. (Image taken from [45]. ©2013 Springer Berlin Heidelberg. Included here by permission.)

4.4.2 Property smoothing

After Eq. (2.27) is solved, vertices can be mapped from spatial space into geometry spectral domain directly with the indices. The chosen properties can be assigned in the spectral domain. As aforementioned, the mean curvature is chosen because the mean curvature normal vector can be obtained by multiplying the Laplacian matrix by the vertex position matrix. The mean curvature is also invariant to Euclidean transformations. However, direct assigning the mean curvature causes stability problems. The embedding is applied on each discrete vertex. A particular vertex usually cannot find the exact matching with other vertices from other surfaces, but has to use the neighbor information in the geometry spectral domain. Based on the definition, the mean curvatures obtained by the mean curvature normal operator use only one ring neighborhood on the mesh. When the mesh is constructed, a noise could be involved during the modeling or reconstruction procedures. Thus the direct mean curvatures will contain a lot of local variance, which will affect the accuracy and stability of the pose analysis in the spectral domain. Therefore they have to be smoothed first.

The smoothing process is done with Laplacian eigenfunctions. As it is discussed in the previous chapters and sections, any function f defined on the surface can be transformed into the frequency domain by projecting it onto the eigenfunctions. The coefficient family $\{c_i\}$ forms the frequency spectrum of f as its counterpart in 1D, which is well known as the Fourier transform. The smoothing is done by applying a low pass filler in the frequency domain and then by transforming filtered coefficients back to the surface function. Fig. 4.6 illustrates the mean curvature reconstruction procedure with different numbers of eigenfunctions. As it is shown, the reconstruction with first 130 eigenfunctions is usually sufficient.

4.5 Automatic skeleton and joint extraction

When all the points on the surface shape are classified and clustered into semantic parts, it enables an automatic skeleton construction with joint identification. Here we adopt the Reeb graph to achieve this goal.

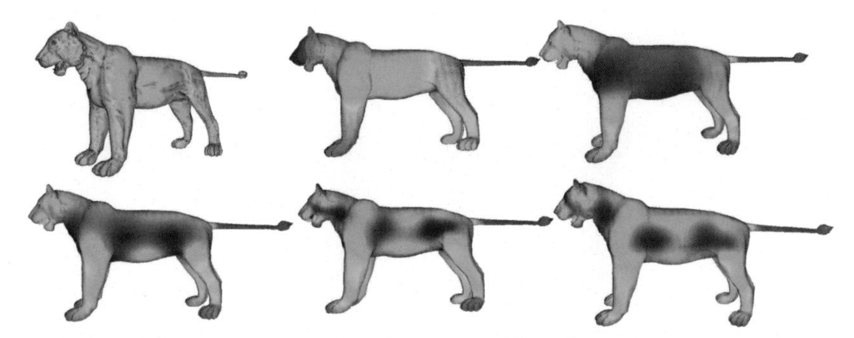

FIGURE 4.6 Mean curvature reconstruction on eigenfunctions of the manifold. From left to right, top to bottom, the first figure is the discrete mean curvatures on the surface, obtained by applying the Laplace–Beltrami operator on the Euclidean embeddings; the remaining ones are reconstructions with the first 6, 20, 50, 100, and 130 eigenfunctions. (Image taken from [45]. ©2013 Springer Berlin Heidelberg. Included here by permission.)

4.5.1 Skeleton extraction

Reuter [118,119] discussed the skeletal representation based on the eigenfunctions. We found that, in the practical data, the intersections of different parts are not stable if the centers of the Reeb graph are directly employed. They may shift away from the semantic locations where they should be. The experiments show that iteratively shrinking the mesh to the center producing smoother results. Our skeleton construction is automatic with two simple steps as demonstrated in Fig. 4.8.

Isocontour shrinking. For each vertex on the mesh, the contour with the same function value of the vertex is traversed and found. Then the vertex is moved to the geometric center of the isocontour. This results in a skeleton-like mesh. Fig. 4.7 illustrates the isocontours of the eigenfunction of the first nonzero eigenvalue.

Skeleton construction. Applying the algorithm in [108] on the shrunk mesh with the original eigenfunction. Because the mesh is shrunk to the skeleton shape, the spatial embedding of the Reeb graph is accurate to become a skeleton.

4.5.2 Joint detection

Based on the changing geometric behaviors of points in the geometry spectral domain, we are able to automatically spot out the joints as long as the deformation around the joints is presented in the given deformation sequence. Fig. 4.5 demonstrates the basic idea of the pose analysis in the geometry spectral domain. Fig. 4.5A is the distribution of the mean curvature maxima on the surface. The larger the value is, the more the surface on that point can bend along relative to the negative direction of the normal at that point. Fig. 4.5B is the distribution of mean curvature minima. It predicts the behavior that surface bends along the positive normal direction. Note that the values on the surface are his-

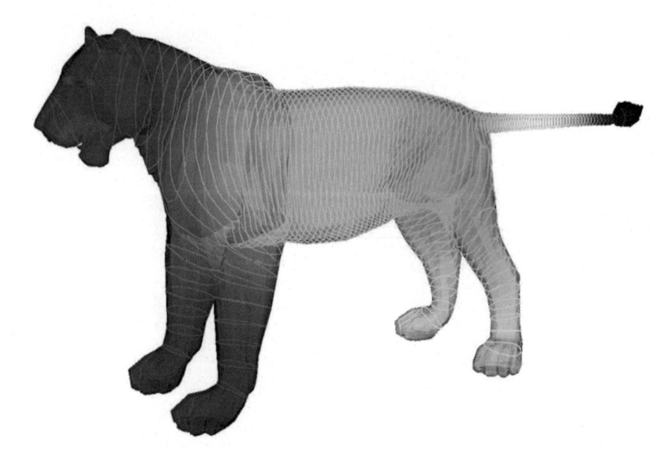

FIGURE 4.7 The isocontours of the eigenfunction of the first nonzero eigenvalue. (Image taken from [42]. Included here by permission.)

FIGURE 4.8 Automatic skeleton generation. From left to right: the first nontrivial eigenfunction of the loin model; shrink mesh based on isocontours; skeleton generated with Reeb graph algorithm; and the embedding of the skeleton within the original model. (Image taken from [45]. ©2013 Springer Berlin Heidelberg. Included here by permission.)

togram equalized. The same color does not mean the same value across different surfaces. Ideally, if a part is always rigid during pose transformation, then the geometry shape never changes. A point on that part has the exact constant mean curvature all the time. Thus the minimum and maximum of mean curvature are equal to each other. On the contrary, if a part varies, then the minima and maxima will fall away from each other. This mean curvature change range is a measurement describing how "rigid" the point and its neighborhood is, which is shown in Fig. 4.5C. The result is very natural. The articulations like neck have different forms under different poses. The parts like nose will not change too much during different poses. Fig. 4.9 shows the complete example.

4.6 Experiments and applications

In this section, we show some experiment results of skeleton and joint extraction and some further applications based on the semantic skeletons. Note that the pose shapes are represented with triangle meshes. In our experiments, we use mesh data sets from SHREC07 and one of Sumner and Popović [135].

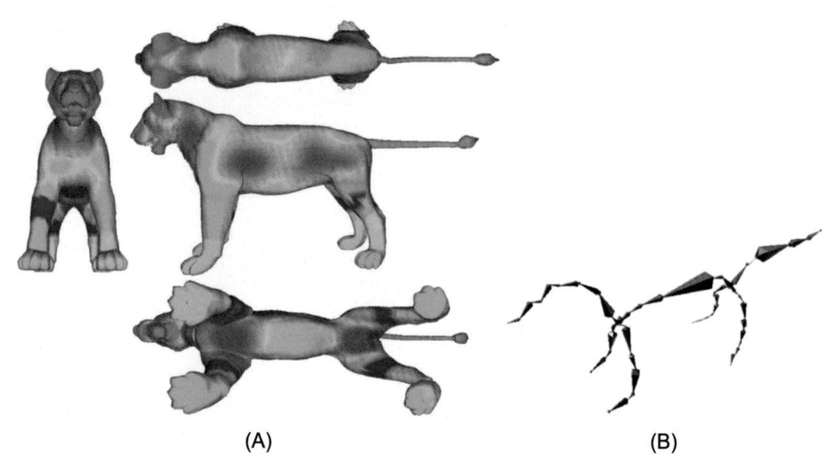

FIGURE 4.9 Mean curvature range distributions on a lion model and the extracted semantic skeleton with joints identified based on the distribution. (Image taken from [45]. ©2013 Springer Berlin Heidelberg. Included here by permission.)

4.6.1 Skeleton and joint extraction

Fig. 4.10 gives an example of the armadillo shape. The main body, especially the chest and the back shell, has no much variance when it casts different poses. Instead, when the armadillo often changes its postures of head, arms, or legs, the neck, shoulder, and waist follow the pose changes. The mean curvature ranges on the surface directly lead to a segmentation, which segments the rigid parts and articulations apart. With the help of the mean curvature ranges, hierarchy graphs can be built as it is described in [21]. Fig. 4.11 shows another example.

4.6.2 Animation

Skeleton driven deformation has extensively studied. It is intuitive to human understanding. Most of poses of creatures are controlled by bones and muscles and then represented by the skin surfaces. The technique of the skeleton-driven deformation and animation is widely used in the animation and gaming industry. The classical pipeline is as follows: first, manually design a skeleton of a mesh surface; second, assign the vertices of that surface to semantic skeletal parts; then deform the mesh along the skeleton. Fortunately, our method automatically classifies semantic parts of surfaces during pose changes and then produces graphs that can be treated as skeletons of meshes. The vertices of the semantically classified surface are automatically associated with skeletal parts with joints identified. Many existing algorithms can be employed to deform and control such a shape with skeletons. Fig. 4.1 has already given an example. The skeletons are learnt from several key frames but can control the shape to cast much more poses than that. Fig. 4.12 also shows some other deformation sequences. These new poses are not any one in the reference frames, but the

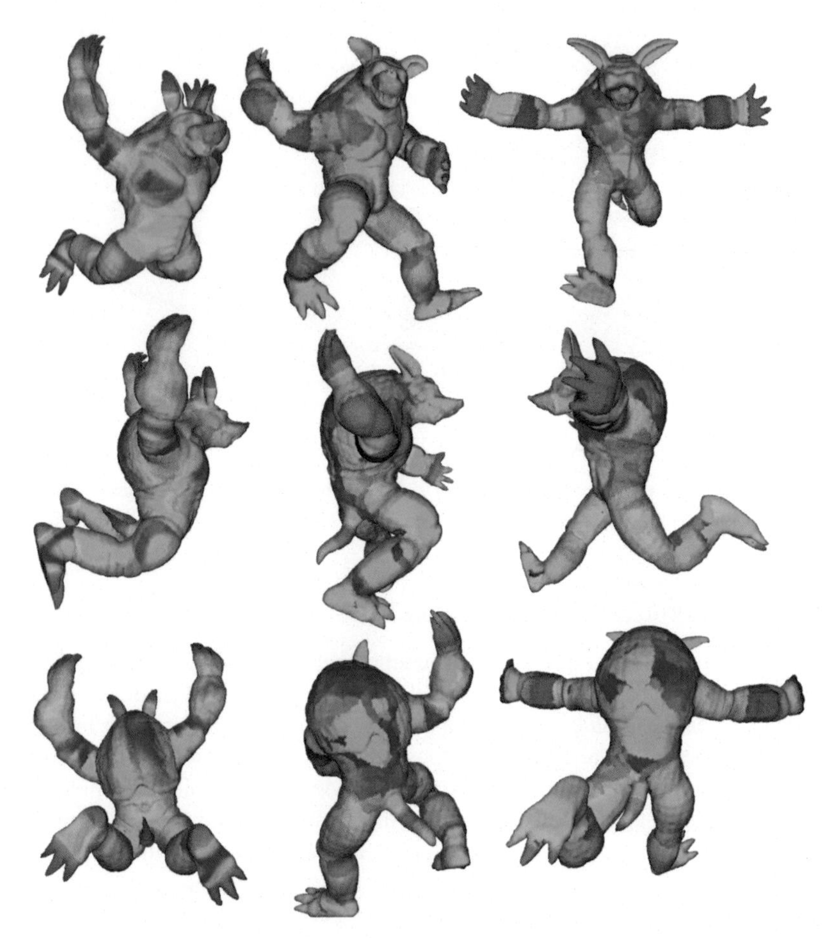

FIGURE 4.10 Mean curvature range distributions on armadillo models. The chest and back shell usually stay rigid, whereas the neck, elbows, and waist vary during pose changes. (Image taken from [45]. ©2013 Springer Berlin Heidelberg. Included here by permission.)

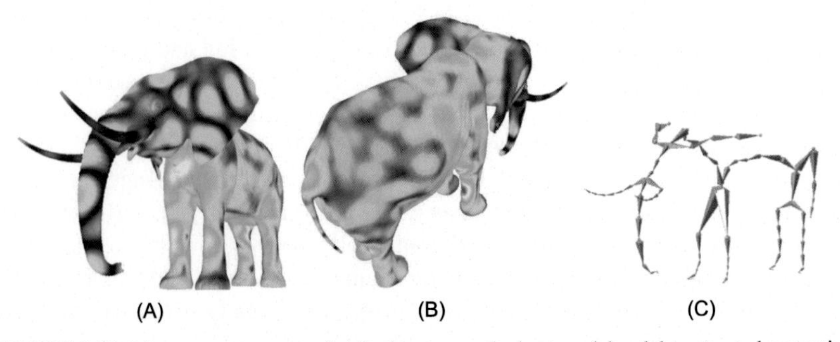

FIGURE 4.11 Mean curvature range distributions on an elephant model and the extracted semantic skeleton with joints identified based on the distribution. (Image taken from [42]. Included here by permission.)

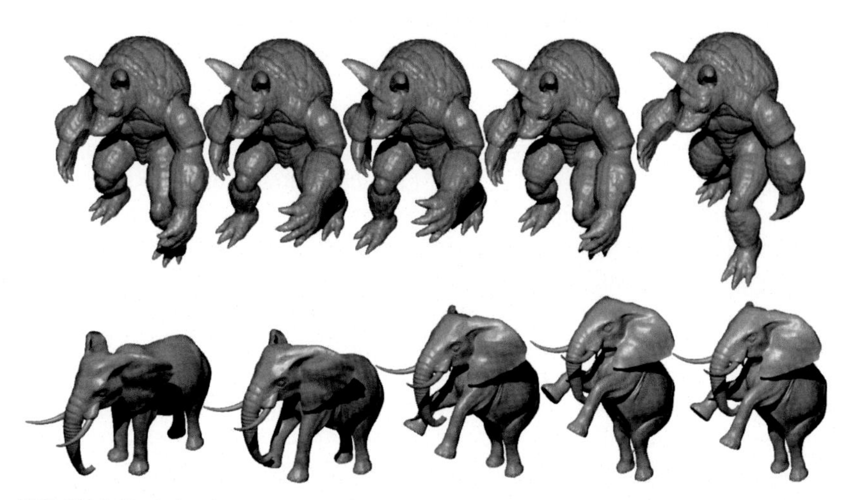

FIGURE 4.12 Animation sequence. With the automatically extracted semantic skeleton, user can edit the pose freely. The animation sequence can be generated among the edited key frames. (Image taken from [45]. ©2013 Springer Berlin Heidelberg. Included here by permission.)

FIGURE 4.13 Motion transform from a lion model to a cat model. (Image taken from [45]. ©2013 Springer Berlin Heidelberg. Included here by permission.)

models can cast some potential possibilities based on the knowledge from existing frames.

4.6.3 Pose transfer

Pose transformation is another popular graphics application. The motivation is obvious. If a pose can be transferred from one shape to another similar shape automatically, a lot of time can be saved by modeling one key shape instead of modeling lots of different shapes, respectively. In our framework, the poses are represented with semantic parts. Two similar shapes will have similar semantic parts and skeletons. A graph or skeleton matching algorithm, such as [137], can find the correspondence between two similar skeletons. After that, a pose driven by a skeleton can be transferred to a similar pose with a corresponded skeleton. Fig. 4.13 demonstrates how running poses are transferred from a lion model to a cat model.

4.7 Summary

Differing from the existing spatial approaches, our method enables to understand the poses in the geometry spectral domain. The geometry spectrum is based on the eigenvalues and eigenfunctions that are defined by the Laplace–Beltrami operator on the surface. The Laplace–Beltrami operator relies only on the metric on the surface, and therefore it is variant to Euclidean translation, rotation, and scaling. It is also invariant to isometric deformations. Thus the eigenvalues, eigenfunctions, and the geometry spectral domain share the invariance. Ideally, every point on a spatial surface should be embedded into the geometry spectral domain only by its geometric meaning. As long as the poses casted by one model are near isometric to each other, they will be reembedded to a uniform surface in the infinite geometry spectrum. In practice the shape spectrum is stable under the near-isometric deformations. For example, the points on the elbow of the model will always be embedded around a common location in the spectral domain, no matter how the model pose changes. The spectrum reflects the intrinsic characteristics of a surface despite varying Euclidean space embeddings.

The discrete setting makes it possible and easy to apply the Laplace–Beltrami operator directly on the surfaces represented by triangle meshes. The continuous Laplacian equation turns into a symmetric generalized sparse matrix eigenproblem. The eigenvalues are kept the same within a finite number, and eigenfunctions are represented with eigenvectors as area-weighted samplings. This also makes the spectral domain invariant to different sampling rates and triangulations.

Our method analyzes data without preprocess like remeshing or registration. It first transforms spatial surfaces into geometry spectral domain. Each point is mapped along with its spatial geometry properties. The properties are smoothed with a low-pass filter defined on the basis of eigenfunctions. In the spectral domain, each point carries a set of properties during the pose variations. It is efficient to classify points on the surface into rigid parts and articulated parts by analyzing the geometric property changes on those points mapped in the geometry spectral domain. The eigenfunction can also provide rich geometric meaning, which leads to an automatic semantic skeleton with joints identified. The experimental results show that the filtered mean curvature range can predict different semantics of parts on the original surface. It may be very useful in motion analysis in computer vision and pattern recognition tasks as well.

Chapter 5

Nonisometric motion analysis by variation of shape spectrum

Contents

In the previous chapters, we have introduced the shape spectrum, which is invariant to different triangulations and isometric deformations. By definition the shape spectrum represents the information of the intrinsic local geometry. A lot of existing approaches and experiments also show that the shape spectrum is stable with noises. The variation of shape spectrum is less studied. In this chapter, we prove that the variation of shape spectrum can be analytically expressed and can be used to analyze nonisometric deformations.

5.1 Nonisometric shape deformation and motion

The spectrum approach started on graphs [96,97,93]. Considering that discrete meshes are also graphs, the Laplacian matrix is defined on vertices and connections, and weights may also be applied. The eigenvalues are defined as the spectra of graphs, and the eigenfunctions are orthogonal bases. This spectrum has a lot similarities with the Fourier transform. The graphs are then projected onto those bases and analyzed in the spectral domain. Karni and Gotsman [58] used the projections of geometry on the eigenfunctions for mesh compression and smoothing. Jain and Zhang [51] extended it for shape registration in the spectral domain. The Laplace spectrum focuses on the connection of graphs, instead of the intrinsic geometry of the manifolds. Only using the connectivity of the graph may lead to highly distorted mappings [158].

The Laplace–Beltrami spectrum also refers to the shape spectrum in this work and was brought to computer graphics to describe shapes [121,79,116, 119]. There are a lot advantages of this spectral approach compared with traditional spatial ones. The spectrum only depends on the intrinsic geometry of a

Spectral Geometry of Shapes. https://doi.org/10.1016/B978-0-12-813842-7.00013-9

manifold. It is invariant to spatial translations, rotations, and scaling. It is also invariant to isometric deformations. On discrete domain, it is well defined on a digital manifold, for example, on triangle meshes. The computing time is affordable. It is invariant to different triangulations. Due to those properties, a lot of shape analysis approaches are based on the Laplace–Beltrami spectrum, including our previous work.

Isometry is a fundamental condition of the shape spectrum. Nonisometry is less studied in this area. By definition the spectrum is not invariant anymore. Reuter et al. [115] discussed that the spectrum is stable to small amount of noise. High-level noise on free-form deformations dramatically changes the spectrum. Recent approaches showed that the shape spectrum can be controlled with a scale function on the Riemannian metric. It is the clue to have the shape spectrum work on general different shapes. Shi et al. [128] discussed that the eigenvalues and eigenfunctions change according to the Riemannian metric of the manifold. The derivative of the eigenvalues can be represented with that of the scale of the Riemannian metric. In [128], eigenfunctions are registered across objects by calculating the Riemannian metric scaling on both shapes. A dense registration is more focused there, and the eigenvalue variation is not studied. Also, the algorithm takes 15 minutes to register two shapes with around thousands vertices, which is not efficient.

In this chapter, we focus on a spectrum alignment for general shapes and also on a computationally affordable discrete algorithm, which can support nonisometric analysis.

5.2 Variation of the eigenvalues and eigenfunctions

In the real-world motion cases, isometry is not usually preserved. Nonisometric deformations result in instability of eigenvalues and dramatic changes of eigenfunctions. In this section, we prove that the eigenvalues are analytic functions of motions.

On a compact closed manifold M with Riemannian metric g, we define a motion as a time-variant positive scale function $\omega(t) : M \longmapsto R^+$ such that $g_{ij}^{\omega} = \omega g_{ij}$ and $d\sigma^{\omega} = \omega d\sigma$, where $\omega(t)$ is nonnegative and continuously differentiable. By definition the weighted Laplace–Beltrami operator becomes

$$\Delta^{g^{\omega}} = \frac{1}{\omega}\Delta^{g}.$$

Consider the ith solution of the weighted eigen problem

$$\Delta^{g^{\omega}} f_i = -\lambda_i f_i, \tag{5.1}$$

rewritten as

$$\Delta^{g} f_i = -\lambda_i \omega f_i, \tag{5.2}$$

where the eigenfunction f_i is normalized as

$$\int_M f_i^2 d\sigma^\omega = 1 \text{ for } i = 0, 1, 2, \ldots, \tag{5.3}$$

and orthogonal to other eigenfunctions:

$$\int_M f_i f_j d\sigma^\omega = 0, \, j \neq i. \tag{5.4}$$

Theorem 1. *The function $\lambda_i(t)$ is piecewise analytic, and, at any regular point, the t-derivative of $\lambda_i(t)$ is given by*

$$\dot{\lambda_i} = -\lambda_i \int_M \dot{\omega} {f_i}^2 d\sigma. \tag{5.5}$$

Proof. The function ω is nonnegative and continuously differentiable, and Δ^g is analytic. We can compute the derivative of the eigenvalue equation (5.2) and get

$$\Delta^g \dot{f_i} = -\dot{\lambda_i}\omega f_i - \lambda_i \dot{\omega} f_i - \lambda_i \omega \dot{f_i}.$$

Then multiplying both sides by f_i and taking the integral on M, we get

$$\int_M f_i \Delta^g \dot{f_i} d\sigma = -\dot{\lambda_i} \int_M \omega {f_i}^2 d\sigma - \lambda_i \int_M \dot{\omega} {f_i}^2 d\sigma - \int_M \dot{f_i} \lambda_i \omega f_i d\sigma,$$

which can be simplified in view of Eqs. (5.2) and (5.3) as

$$\int_M f_i \Delta^g \dot{f_i} d\sigma = -\dot{\lambda_i} - \lambda_i \int_M \dot{\omega} {f_i}^2 d\sigma + \int_M \dot{f_i} \Delta^g f_i d\sigma.$$

Note that M is a closed manifold. According to the divergence theorem, we have

$$\int_M f_i \Delta^g \dot{f_i} d\sigma = -\int_M \nabla \dot{f_i} \cdot \nabla f_i d\sigma = \int_M \dot{f_i} \Delta^g f_i d\sigma,$$

so we get Eq. (5.5). □

For the discrete matrix form, we can get a similar result. Assuming that Ω is a nonnegative continuously differentiable diagonal matrix, we can consider the weighted generalized eigenvalue problem

$$W\mathbf{v}_i = \lambda_i \Omega S \mathbf{v}_i, \tag{5.6}$$

where λ_i and $\mathbf{v}_i$ are the ith corresponding solutions. The eigenvectors can be normalized as

$$< \mathbf{v}_i, \mathbf{v}_i >_{\Omega S} = 1 \text{ for } i = 0, 1, 2, \ldots, \tag{5.7}$$

and orthogonal to each other:

$$< \mathbf{v}_i, \mathbf{v}_j >_{\Omega S} = 0, \ i \neq j. \tag{5.8}$$

Theorem 2. *The function $\lambda_i(t)$ is piecewise analytic, and, at any regular point, the t-derivative of $\lambda_i(t)$ is given by*

$$\dot{\lambda_i} = -\lambda_i \mathbf{v}_i{}^T \dot{\Omega} S \mathbf{v}_i. \tag{5.9}$$

Proof. Computing the derivative of the eigenvalue equation (5.6), we get

$$W\dot{\mathbf{v}}_i = \dot{\lambda_i}\Omega S\mathbf{v}_i + \lambda_i \dot{\Omega} S\mathbf{v}_i + \lambda_i \Omega S\dot{\mathbf{v}}_i.$$

Multiplying by $\mathbf{v}_i^T$ from the left,

$$\mathbf{v}_i^T W\dot{\mathbf{v}}_i = \dot{\lambda_i}\mathbf{v}_i^T \Omega S\mathbf{v}_i + \lambda_i \mathbf{v}_i^T \dot{\Omega} S\mathbf{v}_i + \mathbf{v}_i^T \lambda_i \Omega S\dot{\mathbf{v}}_i,$$

and then simplifying with Eqs. (5.6) and (5.7), we get

$$\mathbf{v}_i^T W\dot{\mathbf{v}}_i = \dot{\lambda_i} + \lambda_i \mathbf{v}_i{}^T \dot{\Omega} S\mathbf{v}_i + \mathbf{v}_i{}^T W^T \dot{\mathbf{v}}_i.$$

Since W is symmetric, we get Eq. (5.9). □

5.3 Computing eigen variation

In the previous chapters and sections, we already discussed the properties of shape spectrum and proved that the eigenvalues can be controlled with a time-variant continuous function. By definition the shape spectrum is invariant to isometric deformations. However, in real cases, the isometry is not guaranteed. For example, in pose deformations the surfaces at the joints are locally scaled, and in heart motions the surface contracts and expands globally. These deformations break the isometry. The former experiments showed that the shape spectrum is stable to these nonisometric deformations and noises. Our eigenvalue variation theorems show that the spectrum is smooth and analytic to a nonisometric local scale deformation. They analytically support aligning the shape spectrum among nonisometric deformations and hence facilitate a registration-free solution for motion analysis.

In this section, we focus on the discrete algorithm to align the shape spectrum among nonisometric deformations. Consider two closed manifolds M and N represented with discrete triangle meshes. Their first k nonzero eigenvalues and eigenvectors are

$$\lambda_{Mi}, \mathbf{v}_{Mi}, \lambda_{Ni}, \text{ and } \mathbf{v}_{Ni} \text{ for } i = 1, 2, \ldots, k.$$

Due to the nonisometry, the first k eigenvalues are not necessary to be aligned. To align the first k eigenvalues of N to those of M, we apply on N a continuous scale diagonal matrix $\Omega(t)$, which is an n by n matrix, where n is the number of vertices on N. The element Ω_{ii} on the diagonal is a scale factor defined on each vertex on N. According to Theorem 2, the derivative of each eigenvalue is analytically expressed by those of Ω_{ii}. Thus the scale matrix Ω introduces an alignment from N to M on the eigenvalues. The rest of this section describes the details to obtain the diagonal matrix Ω numerically.

5.3.1 Linear interpolation

Assume that the eigenvalues of N vary linearly toward those of M. This linear interpolation is represented as

$$\lambda_i(t) = (1-t)\lambda_{Ni} + t\lambda_{Mi}, t \in [0, 1]. \tag{5.10}$$

At the beginning, $t = 0$, and $\lambda_i(0)$ starts as λ_{Ni}, and when t reaches 1, $\lambda_i(1)$ is aligned to λ_{Mi}. At any regular time $t \in [0, 1]$ the derivative is

$$\dot{\lambda}_i(t) = \lambda_{Mi} - \lambda_{Ni}, t \in [0, 1]. \tag{5.11}$$

The derivative of $\lambda_i(t)$ is constant for all t and can be expressed by the derivative of the scale matrix Ω.

5.3.2 Matrix eigenvalue variation

Each diagonal element Ω_{ii} represents a scale factor at vertex i on manifold N; $\Omega(0)$ is the identity matrix on N, and $\Omega(1)$ aligns the first k nonzero eigenvalues of N to those of M. Combining equations (5.9) and (5.11), the derivative of each $\lambda_i(t)$ leads to an equation of Ω:

$$-\lambda_i(t)\mathbf{v}_i(t)^T\dot{\Omega}S\mathbf{v}_i(t) = \lambda_{Mi} - \lambda_{Ni}, t \in [0, 1], \tag{5.12}$$

where S is also the diagonal Voronoi area matrix, and $\mathbf{v}_i(t)$ is the corresponding eigenvector as described in the previous chapters. The diagonal elements of S are defined as the Voronoi area of vertices. Although we have the equation of the time derivative of Ω, it is hidden in the discrete integration and not straightforward to solve. We have to reform the individual integration equation into a linear system. If we extract the diagonals as vectors $\mathbf{v}_\Omega$ and $\mathbf{v}_S$ and employ the Hadamard product, which is an elementwise matrix product

$$A \circ B = C \text{ with } A_{ij} \cdot B_{ij} = C_{ij}, \tag{5.13}$$

then Eq. (5.11) can be rewritten in the linear form

$$(\mathbf{v}_S \circ \mathbf{v}_i \circ \mathbf{v}_i)^T \cdot \mathbf{v}_{\dot{\Omega}} = \frac{\lambda_{Ni} - \lambda_{Mi}}{\lambda_i(t)}, t \in [0, 1]. \tag{5.14}$$

Note that, as the first k eigenvalues are to be aligned, we get k independent equations, which leads to an underdetermined linear system

$$A \cdot \mathbf{v}_{\dot{\Omega}} = \mathbf{b}, \tag{5.15}$$

where A is a row stack of $(\mathbf{v}_S \circ \mathbf{v}_i \circ \mathbf{v}_i)^T$ with k rows,

$$A_{k \times n} = \begin{pmatrix} (\mathbf{v}_S \circ \mathbf{v}_1 \circ \mathbf{v}_1)^T \\ (\mathbf{v}_S \circ \mathbf{v}_2 \circ \mathbf{v}_2)^T \\ \vdots \\ (\mathbf{v}_S \circ \mathbf{v}_k \circ \mathbf{v}_k)^T \end{pmatrix},$$

and $\mathbf{b}$ is a k-dimensional vector with

$$b_i = \frac{\lambda_{Ni} - \lambda_{Mi}}{\lambda_i(t)}, t \in [0, 1].$$

Note that, practically, k is much less than n. For example, on a triangle mesh with 20,000 vertices, only first 50 eigenvalues are aligned. This means that the linear system is undetermined and has no unique solution. More constraints are necessary to provide an optimized solution for the linear system.

5.3.3 Smoothness constraints

In our case, we focus on global smoothness of the scale factors distributed on N. Consider a scalar function $f \in C^2$ on the continuous manifold $< N_c, g >$. The gradient ∇f of f describes the local change of f. For example, if f is a constant function, which is considered as the smoothest distribution, then the gradient ∇f is zero everywhere. The smoothness energy E of f is defined with the total square magnitude of the gradient ∇f on N_c:

$$E = \int_{N_c} \|\nabla f\|^2 d\sigma. \tag{5.16}$$

Note that ∇f is a vector, and the square magnitude is calculated as the dot product

$$\|\nabla f\|^2 = \nabla f \cdot \nabla f. \tag{5.17}$$

Then the integral on N_c becomes

$$E = -\int_{N_c} f \Delta^g f d\sigma. \tag{5.18}$$

At time t, we investigate the scale function $\omega(t)$ and $d\omega|_t$. Then we obtain the following smoothness energy:

$$\begin{aligned} E &= -\int_{N_c} (\omega + d\omega)\Delta^g(\omega + d\omega)d\sigma \\ &= -\int_{N_c} d\omega\Delta^g d\omega d\sigma - 2\int_{N_c} \omega\Delta^g d\omega d\sigma - \int_{N_c} \omega\Delta^g \omega d\sigma. \end{aligned} \tag{5.19}$$

On the discrete triangle mesh N the scale function is a vector $\mathbf{v}_\Omega$, which is the diagonal of matrix Ω. The integral is a matrix product as

$$\begin{aligned} E &=< \mathbf{v}_\Omega + \mathbf{v}_{\dot{\Omega}}, L \cdot (\mathbf{v}_\Omega + \mathbf{v}_{\dot{\Omega}}) >_S \\ &= (\mathbf{v}_\Omega + \mathbf{v}_{\dot{\Omega}})^T \cdot S \cdot L \cdot (\mathbf{v}_\Omega + \mathbf{v}_{\dot{\Omega}}) \\ &= (\mathbf{v}_\Omega + \mathbf{v}_{\dot{\Omega}})^T \cdot W \cdot (\mathbf{v}_\Omega + \mathbf{v}_{\dot{\Omega}}) \\ &= \mathbf{v}_{\dot{\Omega}}^T \cdot W \cdot \mathbf{v}_{\dot{\Omega}} + 2\mathbf{v}_\Omega^T \cdot W \cdot \mathbf{v}_{\dot{\Omega}} + \mathbf{v}_\Omega^T \cdot W \cdot \mathbf{v}_\Omega. \end{aligned} \tag{5.20}$$

Assume that $\mathbf{v}_\Omega$ is known at each time t, and $\mathbf{v}_{\dot{\Omega}}$ is to be solved in Eq. (5.15); $\mathbf{v}_\Omega$ is constant to $\mathbf{v}_{\dot{\Omega}}$, and $\mathbf{v}_{\dot{\Omega}}$ is going to minimize the quadratic smooth energy

$$E_q = \mathbf{v}_{\dot{\Omega}}^T \cdot W \cdot \mathbf{v}_{\dot{\Omega}} + 2\mathbf{c}^T \cdot \mathbf{v}_{\dot{\Omega}}, \tag{5.21}$$

at any time, where $\mathbf{c} = W \cdot \mathbf{v}_\Omega$. To preserve the physical availability, $\mathbf{v}_\Omega$ must be bounded. The scale factor cannot be zero or negative. Furthermore, any point cannot be infinity either. We denote lower upper bounds by $\mathbf{h}_l, \mathbf{h}_u > 0$, where $\mathbf{h}_l$ and $\mathbf{h}_u$ are n-dimensional constant vectors; $\mathbf{v}_{\dot{\Omega}}$ must satisfy

$$\mathbf{h}_l \le \mathbf{v}_\Omega + \mathbf{v}_{\dot{\Omega}} \le \mathbf{h}_u. \tag{5.22}$$

This inequality bound can be written in a matrix form

$$G \cdot \mathbf{v}_{\dot{\Omega}} \le \mathbf{h}, \tag{5.23}$$

where G is the stack of identity matrices,

$$G_{2n \times n} = \begin{pmatrix} -I_{n \times n} \\ I_{n \times n} \end{pmatrix}, \tag{5.24}$$

and $\mathbf{h}$ is a $2n$-dimensional vector,

$$\mathbf{h}_{2n \times 1} = \begin{pmatrix} \mathbf{v}_\Omega - \mathbf{h}_l \\ \mathbf{h}_u - \mathbf{v}_\Omega \end{pmatrix}. \tag{5.25}$$

The linear system (5.15), the smoothness constraint (5.20), and the constant bound (5.23) introduce a quadratic programming problem at each time t. Assuming that the eigenvalues and eigenvectors are known at each time t, the derivative of the scale matrix $\dot{\Omega}$ is the solution of such a quadratic programming problem.

5.3.4 Linear integration

The previous discussions prove that at each time t, the derivative of the scale matrix $\dot{\Omega}$ is the solution of a quadratic programming problem. As an initial state, Ω is the identity matrix as it starts from N itself. The final scale matrix is achieved by the integral

$$\Omega(1) = I + \int_0^1 \dot{\Omega} dt, \tag{5.26}$$

which aligns the first k nonzero eigenvalues from N to M.

This integration is discretely approximated with an iteration. The time interval [0, 1] is divided into K steps. The index of each step is j. Initially, $j = 0$, $\Omega(0) = I$, $\lambda_i(0) = \lambda_{Ni}$, and $\mathbf{v}_i(0) = \mathbf{v}_{Ni}$. To reduce the numerical error, we reinitialize the problem at the beginning of each step $j = 0, 1, \ldots, K$. Instead of aligning λ_{Ni} to λ_{Mi}, we are aligning $\lambda_i(j)$; $\lambda_i(j)$ and $\mathbf{v}_i(j)$ are recalculated with Eq. (5.6) and the current $\Omega(j)$ on N. The diagonal of $\Omega(j)$ is a vector $\mathbf{v}_\Omega(j)$, whereas the diagonal of S is $\mathbf{v}_S$. Then the current quadratic programming problem is constructed with Eqs. (5.15), (5.20), and (5.23) as

$$\begin{aligned}
\underset{\mathbf{v}_{\dot{\Omega}}(j)}{\text{minimize}} \quad & E_q = \mathbf{v}_{\dot{\Omega}}(j)^T \cdot W \cdot \mathbf{v}_{\dot{\Omega}}(j) + 2\mathbf{c}^T \cdot \mathbf{v}_{\dot{\Omega}}(j) \\
\text{subject to} \quad & G \cdot \mathbf{v}_{\dot{\Omega}}(j) \leq \mathbf{h} \text{ (inequality constraint)} \\
& A \cdot \mathbf{v}_{\dot{\Omega}}(j) = \mathbf{b} \text{ (equality constraint)},
\end{aligned}$$

where

$$\begin{aligned}
\mathbf{c} &= W \cdot \mathbf{v}_\Omega(j), \\
G_{2n\times n} &= \begin{pmatrix} -I_{n\times n} \\ I_{n\times n} \end{pmatrix}, \\
\mathbf{h}_{2n\times 1} &= \begin{pmatrix} \mathbf{v}_\Omega(j) - \mathbf{h}_l \\ \mathbf{h}_u - \mathbf{v}_\Omega(j) \end{pmatrix}, \\
A_{k\times n} &= \begin{pmatrix} (\mathbf{v}_S \circ \mathbf{v}_1(j) \circ \mathbf{v}_1(j))^T \\ (\mathbf{v}_S \circ \mathbf{v}_2(j) \circ \mathbf{v}_2(j))^T \\ \vdots \\ (\mathbf{v}_S \circ \mathbf{v}_k(j) \circ \mathbf{v}_k(j))^T \end{pmatrix}, \\
\mathbf{b}_{k\times 1} &= \begin{pmatrix} \frac{\lambda_1(j)-\lambda_{M1}}{\lambda_1(j)} \\ \frac{\lambda_2(j)-\lambda_{M2}}{\lambda_2(j)} \\ \vdots \\ \frac{\lambda_k(j)-\lambda_{Mk}}{\lambda_k(j)} \end{pmatrix}.
\end{aligned} \tag{5.27}$$

Note that, as we reinitialize the problem, λ_{Ni} in Eq. (5.15) is replaced with λ_k at the beginning of each step. If $\dot{\Omega}(j)$ is the solution of the quadratic programming problem, then $\Omega(j+1)$ is approximated with

$$\Omega(j+1) = \Omega(j) + \frac{1}{K-j}\dot{\Omega}(j). \tag{5.28}$$

After K steps, we achieve the desired $\Omega(K)$.

5.3.5 Algorithm pipeline

We summarize the algorithm in Algorithm 1.

Algorithm 1 Eigenvalue alignment.

Input: Closed 2D manifolds N and M, represented triangle meshes, and constant k

Output: Diagonal weight matrix $\Omega(j)$ on N, aligning first k nonzero eigenvalues from N to M

Initialize $\Omega(0) \leftarrow I$, calculate the matrices W and S on N, and $\lambda_{Mi}, \mathbf{v}_{Mi}$, λ_{Ni}, and $\mathbf{v}_{Ni}$ for $i = 1, 2, \ldots, k$

while $j < K$ **do**

Calculate $\lambda_i(j), \mathbf{v}_i(j)$ for $i = 1, 2, \ldots, k$ using Eq. (5.6) with $\Omega(j)$

Construct the quadratic programming problem (5.27)

Solve the quadratic programming problem to get $\dot{\Omega}(j)$

$\Omega(j+1) \leftarrow \Omega(j) + \frac{1}{K-j}\dot{\Omega}(j)$

$j \leftarrow j+1$

end while

5.4 Hands-on experiments and applications to surface motion analysis

Our algorithm is implemented with Python and C++ on a 64-bit Linux platform. The Python libraries Numpy, Scipy and cvxopt are employed for algebra calculations and OpenGL, VTK, and Blender 3D for rendering and visualization. The experiments are conducted on an Intel Celeron 2955U 1.4 GHz laptop with 4 GB RAM. We apply our algorithm on 2D manifolds represented by triangle meshes.

There are typically two kinds of data in our experiments, brains surfaces, and heart left ventricle (LV) motion sequences. They are all extracted from 3D medical images. Each brain and LV surface contains 20,000 and 10,000 vertices, respectively. We first evaluate the computation performance of our algorithm. Besides the vertex number, there are two constants, K iterations, and the first k nonzero eigenvalues to be aligned. According to the algorithm described in the

TABLE 5.1 Performance evaluation. (Table taken from [42]. Included here by permission.)

Manifold	$k = 30, K = 5$	$k = 50, K = 5$	$k = 30, K = 10$	$k = 50, K = 10$
Brain (20,000)	21.2 s	30.6 s	42.6 s	59.1 s
LV (10,000)	8.91 s	14.0 s	18.0 s	27.4 s

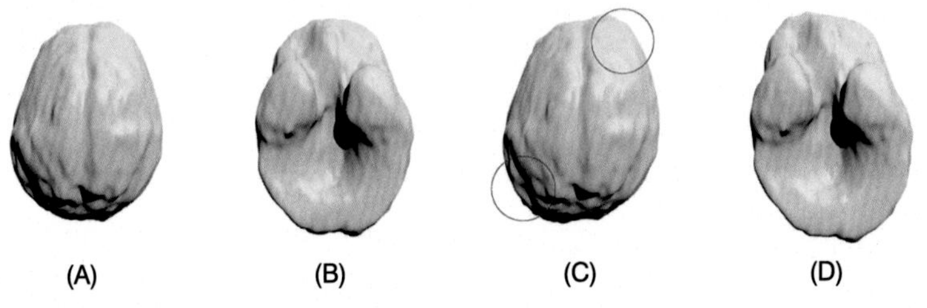

FIGURE 5.1 Synthetic deformation. (A) and (B) are original triangle mesh, which is generate from 3D medical image. (C) and (D) are obtained by manually editing the original surface. The synthetic deformation is local and nonlinear. The locally deformed area are marked with red circle in (C). (Image taken from [42]. Included here by permission.)

previous sections, each iteration is an independent quadratic programming problem. Thus the complexity is linear to the step number K, and k determines how many eigenvalues to be reinitialized at the beginning of each step. The Scipy libraries we employed calculate the eigenvalues by iterations. The complexity is $O(n^2)$ to the number of vertices and linear to k. The average computing time is shown in Table 5.1, which matches the previous analysis. Note that the larger the K, the more accurate the approximation in terms of the linear interpolation. In practice, we found that $K = 5$ is sufficient to get an accurate result and save computing time. Ideally, including more eigenvalues for alignment can be more accurate. However, the numeric eigenvalue calculation is not reliable on higher indexed eigenvalues, which will bring more unsuitability. We usually choose $k = 50$ in our experiments without further mentioning. The computation is quite affordable on a low-profile laptop.

To evaluate our eigenvalue variation algorithm, we synthetically generate some nonisometric deformations. In this case the dense vertex-to-vertex correspondence is known. A reference is chosen as a brain surface, which is extracted from MRI scans. The surface is then deformed manually with local controls, which is nonisometric. The deformation and local controls are shown in Fig. 5.1. Due to the nonisometry, the spectrum is expected to vary, both eigenvalues and eigenfunctions. Table 5.2 shows the eigenvalue variations under the nonisometric deformation. The eigenvalues in the table are normalized by the first nonzero one to remove the scale factor. We can compare how the nonisometry breaks eigenvalue invariabilities from the original brain shape to the synthetic deformation in Table 5.2. Not only the eigenvalues, the eigenfunctions also vary even more dramatically, which is illustrated in Fig. 5.2. We randomly pick the

TABLE 5.2 Eigenvalues alignment on synthetic deformation. (Table taken from [42]. Included here by permission.)

Manifold	λ 2–6	λ 51–55
Synthetic deform.	1.31, 1.36, 2.98, 3.38, 3.69	30.29, 30.69, 31.57, 32.59, 32.94
Brain	1.23, 1.29, 2.93, 3.29, 3.61	29.65, 29.84, 30.96, 31.37, 31.71
Aligned	1.31, 1.36, 2.98, 3.38, 3.69	30.25, 30.66, 31.61, 32.62, 32.87

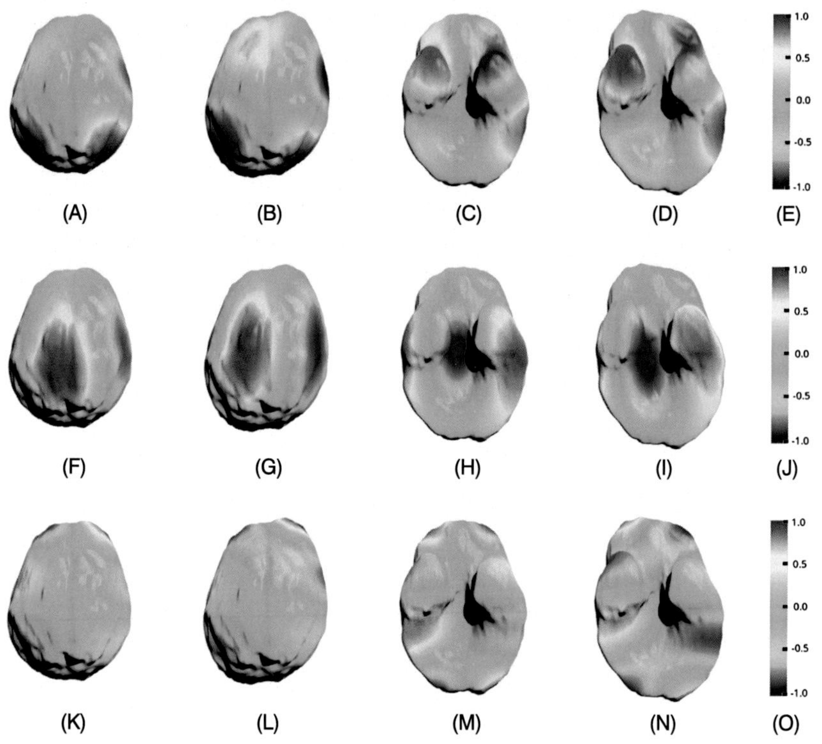

FIGURE 5.2 Synthetic spectrum shifting. The shape spectrum is invariant to isometric deformations. The nonisometric ones break the invariability. We randomly pick the 12th, 14th, and 16th eigenfunctions to show the shifting, represented by each row respectively. The rows of (A) and (C) are the original shape, whereas (B) and (D) are the synthetic deformation. Even small nonisometric deformation introduces noticeable eigenfunction shifting. (Image taken from [42]. Included here by permission.)

12th, 14th, and 16th eigenfunction distributions for comparison, where eigenfunction shifting is more noticeable in the middle range. The eigenfunctions are normalized between -1 and 1. Their values are expressed by color maps, where red means larger value, blue means smaller ones, and green means zero. The patterns of the eigenfunctions shift around. They may not represent the corresponding geometry across nonisometric manifolds. In the worst case, the

TABLE 5.3 Average normalized eigenvalue errors before and after alignment. (Table taken from [42]. Included here by permission.)

	Synthetic		Brain		LV	
	first 50	first 100	first 50	first 100	first 50	first 100
Before	8.93%	9.00%	6.55%	6.61%	25.9%	24.5%
After	0.0186%	0.0187%	0.202%	0.203%	0.290%	0.286%

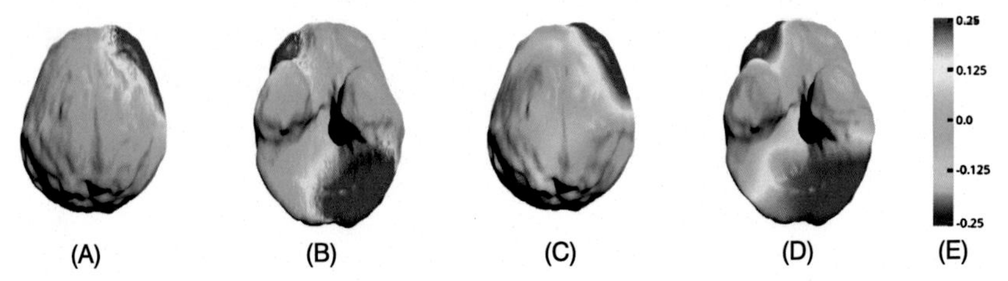

FIGURE 5.3 Synthetic deformation log ratio. The scale function from the eigenvalue alignment algorithm is evaluated with the synthetic ground truth. With the synthetic deformation, the dense vertex-to-vertex correspondence is known. The vertexwise scale function is calculated with the Voronoi area ratio after and before the deformation, demonstrated in (A) and (B). To make it linear for comparison, the log operation is employed. (C) and (D) represent the scale function from the eigenvalue alignment. Our algorithm accurately recovers the local deformation without any preinformation but shape spectra. (Image taken from [42]. Included here by permission.)

topology of the eigenfunction distribution may also change. With the synthetic deformation, our spectrum alignment is applied on the first 50 nonzero eigenvalues, and a scale function is obtained. We compare the eigenvalues, on the original brain and after alignment, in Table 5.2. After applying the spectrum alignment algorithm, the eigenvalues are aligned perfectly. Furthermore, the higher indexed eigenvalues are also aligned even if they are not in the linear system constraint. We define the relative error of eigenvalues as

$$er = \frac{|\lambda_i^{\text{reference}} - \lambda_i|}{\lambda_i^{\text{reference}}}.$$

Table 5.3 records the accuracy of the spectrum alignment algorithm in percentage on the eigenvalues. Both ranges, within the first 50 constraints and first 100, are measured. The alignment reduces the eigenvalue error by two orders of magnitude. Fig. 5.3 demonstrates the scale function distributions on the original surface. The color represents the log of the scales. Red means dilating, blue contraction, and green no distortion. (A) and (B) are ground truth by calculating the vertex-to-vertex Voronoi area distortions on the synthetic surface. (C) and (D) are the result of the spectrum alignment. It is clear that the spectrum alignment predicts the local deformation precisely, with spectrum information only. Please note that the resulting scale function is much smoother than that of ground truth, and they have slight difference, because we are using smoothness constraints to

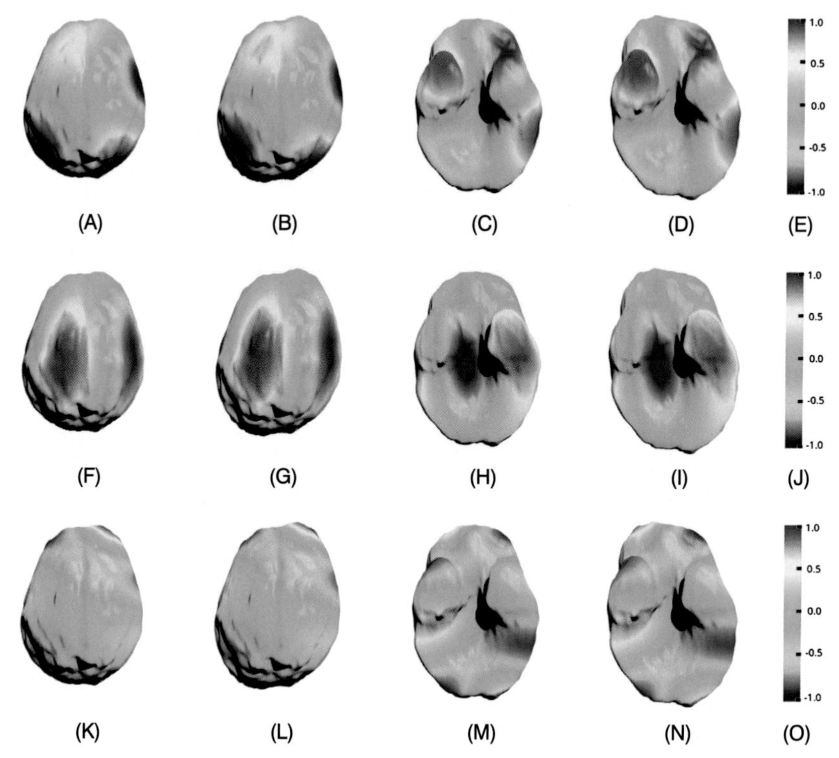

FIGURE 5.4 Synthetic spectrum alignment. The eigenvalues are aligned from the original shape to the synthetic deformation. The eigenfunctions are aligned as well. The 12th, 14th, and 16th eigenfunctions are represented by each row respectively. Those eigenfunctions are more consistent after eigenvalue alignment over deformations. (Image taken from [42]. Included here by permission.)

solve the linear system. The next step is investigating how the eigenvalue distributions change under the spectrum alignment. We pick the same eigenfunctions, the 12th, 14th, and 16th, after alignment for comparison in Fig. 5.4. Compared with the unaligned eigenfunctions in Fig. 5.1, those ones after the spectrum alignment are more responsible. Especially, the 16th eigenfunction on the top has not only shifting but also topology change on the top. The blue part is a connected strip shape on the deformed shape but separates on the original brain, which is also corrected with the spectrum alignment in Fig. 5.4. The synthetic deformation is aligned in the shape spectrum.

We move the next step to real clinical data. Brain surface registration is a fundamental research in image analysis and vision. The brain data in our experiment are extracted from MRI scans with marching cube and resampled with 20,000 vertices for each shape. The shapes from different persons are usually nonisometric to each other. The original shape spectrum fails in this case. In Table 5.3 the average eigenvalue error is 6.55% among different brain surfaces.

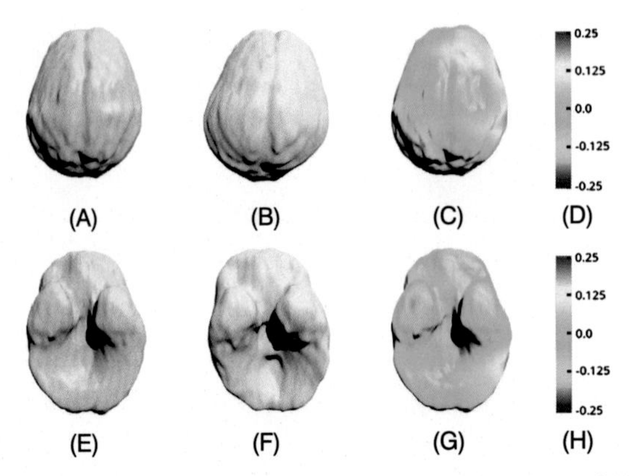

FIGURE 5.5 Different shapes are aligned with a scale function. The column of (A) is the reference shape, (B) is the target one to be aligned, and (C) is the scale function distribution on the reference shape. The color represents the log values of the scale factors. (Image taken from [42]. Included here by permission.)

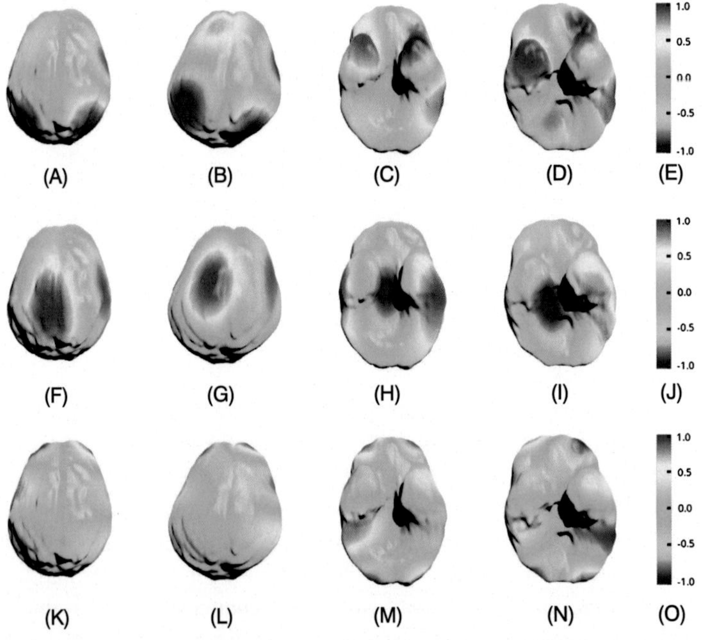

FIGURE 5.6 Brain spectrum shifting. The two brains shapes are from different persons, which are not isometric to each other. The rows show the 12th, 14th, and 16th eigenfunction distributions on the two brains. The columns of (A) and (C) represent one brain, and (B) and (D) the other. The eigenfunctions shift due to the nonisometry. (Image taken from [42]. Included here by permission.)

Two brains samples are demonstrated side by side in Fig. 5.5. It is obvious that they are not isometric even up to a global scale factor. Fig. 5.6 shows how the eigenfunction distributions change among different brains. Here the 12th, 14th,

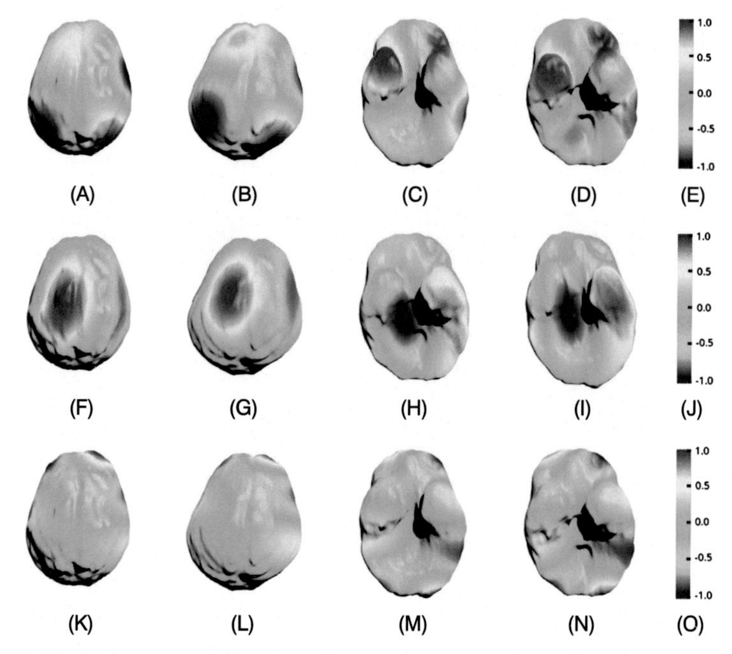

FIGURE 5.7 Brain spectrum alignment. The spectra of the two brains are aligned with a scale function. The rows show the 12th, 14th, and 16th eigenfunction distributions respectively on the two brains. The columns of (A) and (C) represent one brain and (B) and (D) the other. The eigenfunctions are aligned as well. (Image taken from [42]. Included here by permission.)

and 16th eigenfunctions are chosen. The patterns, for example, maxima and minima, are shifting around, and some of them are even missing. The spectra of these two brains can be aligned with a scale function. Such a scale function is illustrated in Fig. 5.5, whose log values are mapped with color. Intuitively, the red and yellow areas expand themselves, whereas the blue ones contract. Then the geometry on the left brain is deformed to the right one. Table 5.3 shows that the spectrum alignment reduces the eigenvalue error from 6.55% to 0.202%. The eigenfunctions are also aligned as they are geometrically more similar than the original case; see Fig. 5.7. Comparing Figs. 5.7 and 5.6, a great improvement is obtained after spectrum alignment, in terms of those maxima, minima, and transition edges on the geometry. Another interesting case to study is LV motion in heart. LV changes its shape within cycles by contracting and dilating the muscles. This motion results in changes of the surface geometry. Fig. 5.8 shows a sequence of eight samples during an LV motion. It can been seen that the geometry changes dramatically. The middle sample, the first one in row 3, is chosen as a reference shape, and its spectrum is aligned to all the other shapes. The eigenvalue errors shown in Table 5.3 verify that the large geometry changes as the average error is 25.9% and the spectrum alignment brings it down to 0.290%. Figs. 5.9 and 5.10 demonstrate the eigenfunction distributions before and after

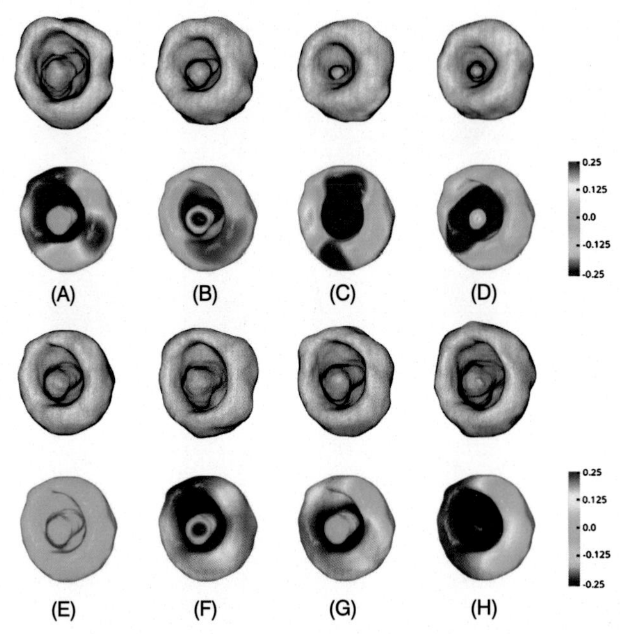

FIGURE 5.8 LV motion represented by scale functions. Although the LV motion introduces a sequence of nonisometric deformations, the spectra of those deformations can still be aligned with scale functions. Each deformation is then represented with a scale function on the reference frame. (Image taken from [42]. Included here by permission.)

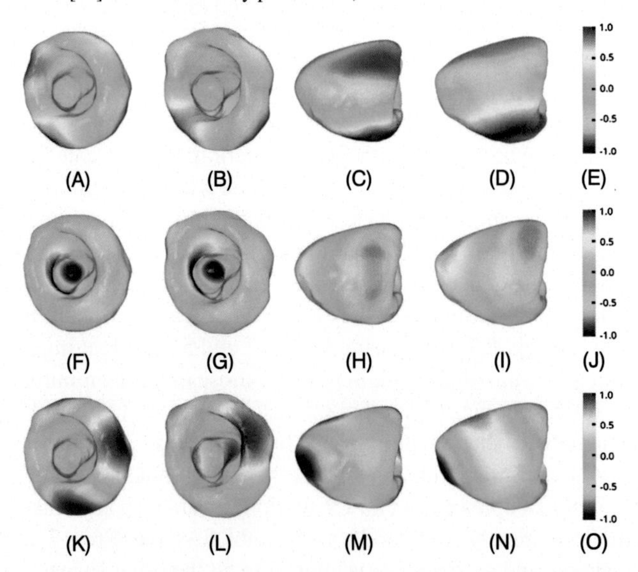

FIGURE 5.9 LV spectrum shifting during motion. The local parts of an LV contract and expand. Those deformations are usually not isometric. The rows show the 8th, 12th, and 14th eigenfunction shiftings, respectively. The columns of (A) and (C) represent one time frame in the LV motion, and (B) and (D) the other. (Image taken from [42]. Included here by permission.)

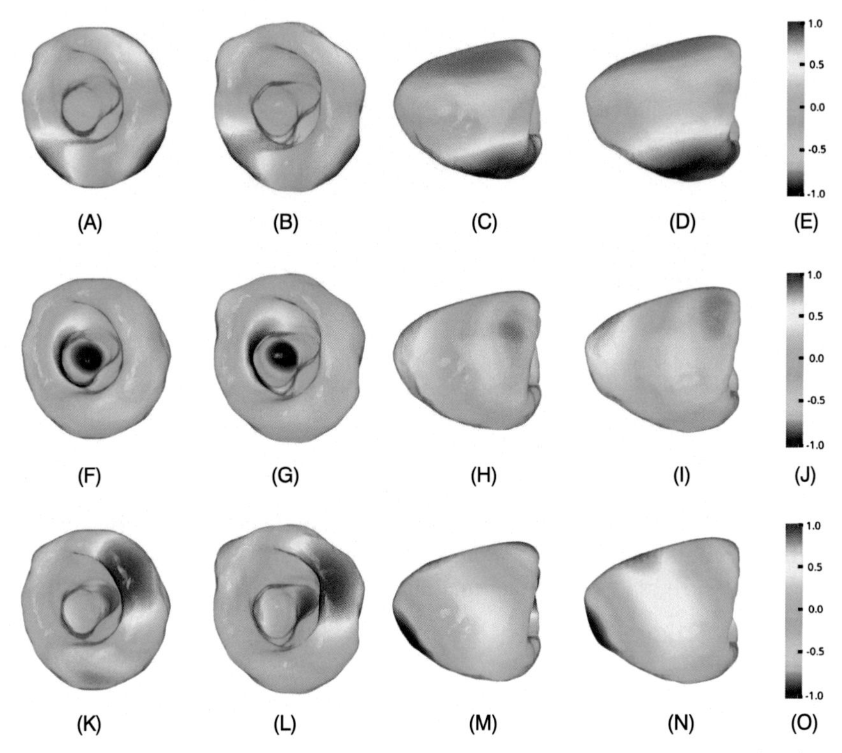

FIGURE 5.10 LV motion spectrum alignment. The LV motion introduces a sequence of nonisometric deformations. The spectra of each time frame can be aligned with a scale function. Both eigenvalues and eigenfunctions are aligned during the motion. The rows show the 8th, 12th, and 14th eigenfunction shiftings, respectively. The columns of (A) and (C) represent one time frame in the LV motion, and (B) and (D) the other. (Image taken from [42]. Included here by permission.)

the spectrum alignment. The results are similar to the previous synthetic shape and brain surfaces. Whereas the eigenvalues are aligned among the shapes, the eigenfunctions are also aligned according to the geometry. The spectrum alignment results in a scale function to each time frame as illustrated in Fig. 5.8. Red, blue, and green colors indicate dilation, contraction, and no scaling, respectively. For example, from the reference shape to the first frame, the interior has to expand, and to the third frame, it is necessary to contract instead. In this way the spacial geometry deformations turn into scale function distributions. Note that the scale functions are globally smooth and predict local deformations. Another potential application would be abnormality visualizations and diagnosis. For example, cardiomyopathy is the main cardiac disease, which affects the wall thickness and its functionality. This disease can be detected from the abnormal motion of LV [138]. The variations are usually weaker on the myopathy parts than those on the normal parts. In this work the scale function illustrates the contractions and expansions of the local part. For rendering, we cut interior wall

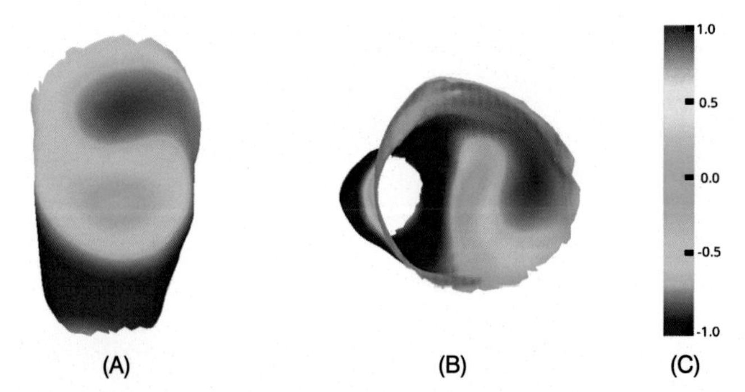

FIGURE 5.11 LV abnormality on the interior wall. Blue color indicates contractions at this time frame. The major parts of the interior wall contract normally. Some myopathy ones have much less or no deformations, which are colored with green. (Image taken from [42]. Included here by permission.)

of the LV, as shown in Fig. 5.11. It is during the contraction phase. Most of the surface is contracting, which is indicated by blue color. There is an abnormal patch that receives less or no deformations.

5.5 Summary

In this chapter, we have introduced spectrum variation theorems for general shapes. A shape is represented with a closed 2D manifold with Riemannian metric. The Laplace–Beltrami spectrum is defined on the intrinsic geometry of the manifold, which is invariant to rigid operations and isometric deformations. In real cases, isometry is hard to preserve. Even small nonisometric deformation causes spectrum variations. We prove that the eigenvalues of the spectrum are an analytic function of a scale function applied on the Riemannian metric. The derivative of each eigenvalue is an integral of the derivative of the scale function. The theorem applies both continuous analytic and discrete cases.

We have also developed an algorithm to align the shape spectra on discrete shapes represented with triangle meshes. Given two closed triangle meshes, the spectra can be aligned from one to another with a scale function defined on each vertex. The alignment is expressed as a linear interpolation of eigenvalues. The interpolation is then decomposed into discrete iterations. In each step, a quadratic programming problem is constructed with the spectrum variation theorem and smoothness energy constraint. The derivative of the scale function is a solution of such a problem. The final scale function is approximated with integral of the derivative from each step.

Our experiments verify the spectrum variation theorem and illustrate the accuracy and efficiency of the alignment algorithm on general shapes and their motions to shape motion analysis.

Chapter 6

Nonisometric registration using eigenvectors and eigenvalues

Contents

In this chapter, we present a novel surface registration technique using the spectrum of the shapes, which can facilitate accurate localization and visualization of nonisometric deformations of the surfaces. To register two surfaces, we map both eigenvalues and eigenvectors of the Laplace–Beltrami of the shapes through optimizing an energy function. The function is defined by the integration of a smoothness term to align the eigenvalues and a distance term between the eigenvectors at feature points to align the eigenvectors. The feature points are generated using the static points of certain eigenvectors of the surfaces. Using both the eigenvalues and eigenvectors on these feature points, we considerably improve the computational efficiency without losing the accuracy in comparison to the approaches that use the eigenvectors for all vertices. In our technique the variation of the shape is expressed using a scale function defined at each vertex. Consequently, the total energy function to align the two given surfaces can be defined using the linear interpolation of the scale function derivatives. Through the optimization of the energy function, the scale function can be solved, and the alignment is achieved. After the alignment, the eigenvectors can be employed to calculate the point-to-point correspondence of the surfaces. Therefore

Spectral Geometry of Shapes. https://doi.org/10.1016/B978-0-12-813842-7.00014-0

the proposed method can accurately define the displacement of the vertices. We evaluate our method by conducting experiments on synthetic and real data using hippocampus, heart, and hand models. We also compare our method with non-rigid ICP and a similar spectrum-based method. These experiments demonstrate the advantages and accuracy of our method.

6.1 Introduction to nonisometric shape registration

Shape registration is one of the important research topics for scientific visualization, computer vision, and shape analysis. In biomedical area, its application ranges from analysis of cardiac deformations [5] to brain structure deformations caused by diseases such as epilepsy [100] or Alzheimer [47]. Since the deformations of most organs such as heart or brain structures are nonisometric, finding the correspondence between the shapes before and after deformation is very challenging for diagnosis purposes.

Traditional landmark-based methods usually detect relevant corresponding points or curves in two shapes, that is, landmarking is essential in many shape registration and mapping applications [48,99,104,53]. There are two drawbacks of this type methods. First, due to the shape complexity of organs, these methods require labor-intensive human intervention when done manually or error-prone if conducted automatically through spatial detection [155]. Secondly, in many situations, there exist no salient spatial landmarks in the nonisometric deformations, for example, in left ventricle of heart or brain hippocampus.

The shape spectrum is another method to represent the shape. There is a powerful tool called the Laplace–Beltrami operator, which can analyze the intrinsic property of the shape. Employing this operator, Reuter [116] and Lévy [79] defined a shape spectrum approach with the Laplace–Beltrami operator on a manifold and employed the eigenvalues and eigenvectors as a global shape descriptor [115,117]. The eigenvectors are orthogonal basis functions; therefore the shape can be projected to the orthogonal bases and then analyzed and reconstructed using these bases [129]. As the geometry changes, the spectrum of the shape changes as well. Some studies employed the spectrum of this operator to classify, register, and differentiate shapes [70,66,71,44]. However, the spectrum through these methods can only show the global difference between shapes and cannot map and quantify the nonisometric shape differences due to the lack of nonisometric registration with spectrum. Hamidian et al. [36,43] proposed an alignment method through the eigenvalues; however, the point-to-point correspondence cannot be determined. Cosmo et al. [24] employed a similar method to align eigenvalues. After the eigenvalue alignment, they used the similarity between the eigenvectors to find the corresponding points. This method is applicable to cases in which the eigenvectors turn out to be similar after aligning the eigenvalues. However, in the cases of dramatic deformations like heart beating, just aligning the eigenvalues cannot make eigenvectors very similar. Therefore the correspondence among deformations cannot be correctly identified. In this kind of dramatic deformation, eigenvectors need to be involved in the alignment process to find the correct corresponding points. Shi et al. [130] used the difference between the eigenvectors of two surfaces to generate a conformal mapping, but the method is computationally expensive. Although many promising techniques were developed, there is still a lack of a method that can generate the correspondence between points for nonisometric shape structure change in a timely efficient fashion.

In this chapter, we focus on a method based on spectrum alignment of the nonisometrically deformed surfaces using both eigenvalue and eigenvector variations to find the correspondence and map the nonisometric deformations. To search for the alignment, we use a scale function on the surface that deforms one surface to a targeted one. Compared to the traditional approaches through the experiments, our method can accurately and automatically map and localize the point-to-point nonisometric deformations in addition to global difference of the shapes. Because the spectrum of shape only depends on the intrinsic geometry, our method is invariant to spatial translation, rotation, scaling, and isometric deformation. Furthermore, our method is computationally efficient and takes considerably less time to execute compared with existing methods [130].

6.1.1 Related work

By definition the shape spectrum represents the information of intrinsic local geometry. It is invariant to isometric deformations and different triangulations. Reuter et al. [116] defined the spectrum of the Laplace–Beltrami operator of a shape as the signature or fingerprint of the shape. Rustamov [121] employed the spectrum of this operator for shape clustering and classification purposes. Lévy [79] employed the theory of *stationary waves* to study the behavior of eigenvectors and the static points of an eigenvector. These points correspond to the locations that do not move in the theory of stationary waves. This study showed that the static points are strongly linked to the geometry of a shape, and locations of these points change when changing the geometry. As these points are extracted from eigenvectors, they are invariant to isometric deformations of the shape. Thus these points can be employed as the feature points to describe the geometry of the shape. Reuter et al. [119,118] employed these points, together with the domains generated by these static points, as topological features to segment and register different parts of the shapes. However, the deformation of the shapes is restricted to be isometric in these studies. In reality, many deformations, such as heart motion, brain development, and so on, are not isometric. Hence, applying geometric spectrum methods for analyzing nonisometric deformation and registration is very challenging. Some recent works [128,130] showed that the shape spectrum can be controlled with a scale function on the Riemannian metric. Shi et al. [128] discussed that the eigenvalues and eigenvectors change according to the Riemannian metric of a manifold. Later, Shi et al. [130] employed this metric to measure the difference between the eigenvectors of two surfaces to generate a conformal mapping between them. To this end, they minimized the difference between surfaces in the Laplace–Beltrami embedding space using an optimization approach. This work focused on the eigenvector variation, but the eigenvalue variation was not investigated, and the method is very computationally expensive. For instance, to map two hippocampal surfaces with 1000 faces, the procedure took around 20 minutes on a computer with a 2.6 GHz Intel Xeon CPU and approximately 60 MB memory consumption. Instead, Hamidian et al. presented a method to align two surfaces by mapping their eigenvalues [36,43]. This method provided a deformation matrix showing the deformation of the initial surface to the target one but did not generate a point-to-point correspondence mapping of the vertices.

There is another work that employs the shape spectrum to match shapes. Rodolà et al. [120] proposed a method based on the Laplace–Beltrami eigenvectors for computing partial functional correspondence between nonrigid shapes that have isometric deformation. Litany et al. [83] extended this study to match partial shapes that undergo topological noise and nonisometric deformation within the same framework. There are some limitations for this method. The main limitation lies in its reliance on good local features to drive the matching process. There are recent advances in the field of spectral shape analysis closely related to the proposed approach. For instance, Kovnatsky et al. [67] showed how to modify (align) the eigenvectors of the Laplace–Beltrami operator to match nonisometric shapes. Ovsjanikov et al. [107] proposed a spectral method for shape matching based on finding an alignment between eigenvectors under a set of linear constraints. Later, they [106] presented a method for finding functional correspondence between manifolds based on the geometric matrix completion framework [68]. In [106,122,156], visualizing shape deformations based on a spectral representation of the correspondence was shown. However, the key difference between the methods mentioned and our proposed approach lies in the fact that our method is using both eigenvalues and eigenvectors to align two manifolds versus employing only the eigenvectors. Also, our method extracts feature points from eigenvectors, instead of using all the points, and employs them to align two surfaces without losing accuracy.

In this chapter, we present a novel method that can align two surfaces and visualize the corresponding points through the variation of geometric spectrum. This is achieved by mapping eigenvalues and certain feature points extracted from the eigenvectors of two surfaces. Given two triangle meshes, the spectra can be varied from one to another with a scale function defined on each vertex. To compute the alignment, we aim to minimize an energy function, which is the integration of a smoothness term for aligning the eigenvalues and a distance term describing the distance between the corresponding feature points. Optimizing this energy function is a quadratic programming problem, which can be solved using an iterative method. Furthermore, we assume that the variation of an eigenvalue is expressed as a linear interpolation of eigenvalues of the two surfaces. The derivative of the scale function is the solution of such a problem. Therefore the final scale function can be computed by an integral of the derivatives from each step. Subsequently, the scale function can describe the mapped surface eigenvectors that can be employed to find the point-to-point correspondence. Our major contributions in this work can be summarized as follows:

- **We present a spectrum alignment algorithm using eigenvalue and eigenvector variations for 3D surfaces and supporting nonisometric global and local deformation analysis.** In the discrete domain the variation of eigenvalues and eigenvectors in terms of the scale function can be presented as matrices. Employing these matrices, together with the smoothness function to align the eigenvalues and a distance function to align the feature points extracted from eigenvectors, a linear system can be defined. By solving this system the eigenvalues and eigenvectors are aligned, and the corresponding points of the surfaces can be determined.
- **Feature points automatically extracted from the eigenvectors of the surfaces, along with the defined distance between the corresponding feature points, can lead to an improved correspondence with considerably reduced computational cost.** Because our method aligns both eigenvalues and eigenvectors at the same time, a limited number of feature points for the eigenvectors are sufficient to warrant the alignment. This helps to improve the accuracy and considerably reduce the computational time. These feature points are proven to be highly related to the geometry of the shape, and they change when deforming the shape.
- **Our developed system demonstrates the accuracy and efficiency of the spectral variation and registration algorithm on visualization of nonisometrically deformed shapes.** Applications to biomedical imaging problems show that it is a viable solution for morphometric analysis and visualization in biomedical applications and clinical diagnoses.

6.2 Surface registration using spectral optimization

In this chapter, we again employ the Laplace–Beltrami operator to compute the geometric spectrum of a manifold. Let $f_1 \in C^2$ be a real function on a Riemannian manifold M. The Laplace–Beltrami operator $\triangle$ is defined as $\triangle f_1 = \nabla \cdot (\nabla f_1)$, where ∇f_1 is the gradient of f_1, and $\nabla \cdot$ is the divergence on the manifold M. The eigen-system for this equation is defined as $\triangle f = -\lambda f$, where the family solution $\{\lambda_i\}$ is a real nonnegative scalar and results in the corresponding real family functions of $\{f_i\}$ for $i = 0, 1, 2, \ldots$. To solve these differential equations, a discrete differential operator is employed [94]. In this framework, the Voronoi region for the vertices of a triangle mesh is used to construct the Laplacian–Beltrami matrix as

$$L_{ij} = \begin{cases} -\frac{\cot\alpha_{ij}+\cot\beta_{ij}}{2A_i} & \text{if } i, j \text{ are adjacent,} \\ \sum_k \frac{\cot\alpha_{ik}+\cot\beta_{ik}}{2A_i} & \text{if } i = j, \\ 0 & \text{otherwise,} \end{cases} \tag{6.1}$$

where α_{ij} and β_{ij} are the angles opposite to the edge in the two triangles sharing the edges i and j, A_i is the area of Voronoi region at vertex i, and k are the indices of triangles within 1-ring neighborhood of the vertex i. Therefore the eigenequation becomes $\mathbf{L}f = \lambda f$, where f is an n-dimensional vector for each λ in which n is the number of vertices of a manifold. To solve this equation, we use a sparse matrix $\mathbf{W}$ and a diagonal matrix $\mathbf{S}$ such that

$$W_{ij} = \begin{cases} -\frac{\cot\alpha_{ij}+\cot\beta_{ij}}{2} & \text{if } i, j \text{ are adjacent,} \\ \sum_k \frac{\cot\alpha_{ik}+\cot\beta_{ik}}{2} & \text{if } i = j, \\ 0 & \text{otherwise,} \end{cases}$$

and $S_{ii} = A_i$. Thus the Laplace Matrix $\mathbf{L}$ can be written as $\mathbf{L} = \mathbf{S}^{-1}\mathbf{W}$, and the eigenequation can be presented as

$$\mathbf{W} f_n = \lambda_n \mathbf{S} f_n, \tag{6.2}$$

where f_n and λ_n are the nth eigenvector and eigenvalue, respectively. The eigenvectors for different eigenvalues are orthogonal in terms of the $\mathbf{S}$ dot product. Using this concept, the following embedding $\mathbf{I}_M : M \to R^\infty$ is proposed [121]:

$$\mathbf{I}_M^\Phi = \left(\frac{f_1(x)}{\sqrt{\lambda_1}}, \frac{f_2(x)}{\sqrt{\lambda_2}}, \ldots, \frac{f_n(x)}{\sqrt{\lambda_n}}\right), \quad x \in M, \tag{6.3}$$

where $\Phi = \{f_0, f_1, f_2, \ldots\}$. Two shapes can be aligned by finding the proper mapping between the eigenvector embeddings after solving the sign ambiguity. Considering that for each eigenvalue, there is an eigenvector of size n, mapping eigenvectors of two surfaces for all the vertices is time-consuming. Therefore we propose to use the eigenvector values for certain feature points to map the shapes.

6.2.1 Calculating the feature points

Using the spectrum of the Laplace–Beltrami operator, Lévy [79] employed the theory of stationary waves to model the shape. The spectrum contains a lot of information about the shape, which can therefore be used for matching and mapping among different shapes. Looking closely to the eigenvectors, it shows that the nth eigenvector can have at most n nodal domains. The nodal domains are

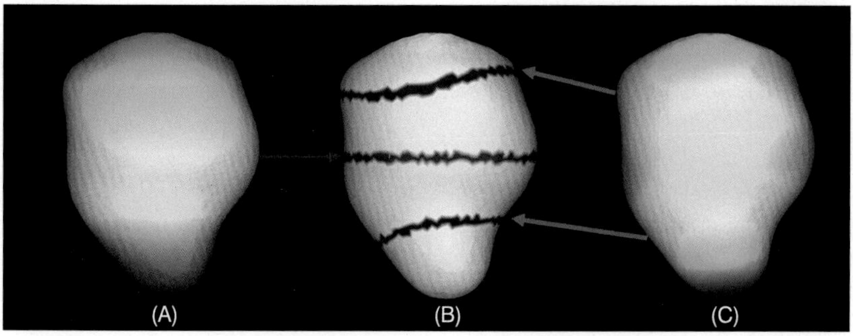

(A) (B) (C)

FIGURE 6.1 (A) The eigenvector corresponds to the first nonzero eigenvalue. (B) The three nodal sets. The red set shows the static points for eigenvector corresponding to the first nonzero eigenvalue. The blue sets show the static points for eigenvectors corresponding to the second nonzero eigenvalue. (C) The eigenvector corresponds to the second nonzero eigenvalue. (Image taken from [37]. ©2019 IEEE. Included here by permission.)

the partitions of the surface that have the same sign. In this work, we are interested in the points, called *nodal sets*, that are the static points between two nodal domains. In other words, the nodal sets separate the nodal domains. These nodes are the still zones in the theory of stationary waves. Lévy [79] showed that these points are strongly linked to the geometry of the shapes. In our method, we use the nodal sets of certain eigenvectors as the feature points to map the eigenvectors of two shapes. As mentioned before, the nth eigenvector has at most $n - 1$ nodal sets that partition the n nodal domains. We use this concept and employ the nodal sets of the eigenvectors corresponding to the first nonzero eigenvalue as the first set of feature points. Fig. 6.1A shows the first nonzero eigenvector for a sample left ventricle of heart, and the red set of points in Fig. 6.1B shows this first feature set. The second feature sets are the nodal sets for the eigenvector corresponding to the second or third nonzero eigenvalue that are parallel to the first set of feature points. We use the parallel points to get exclusive sets of points for mapping the eigenvector of two shapes. Fig. 6.1C shows the eigenvector corresponding to the second nonzero eigenvector, and the blue sets of points on Fig. 6.1B present the second set of feature points. This approach can provide us three sets of points that are used for matching the eigenvectors of different shapes. When needed, more nodal sets can be used. Using these feature points for mapping the eigenvectors, instead of using all of the points, reduces the computational time considerably.

6.2.2 Spectral registration using eigenvector and eigenvalue

To register two surfaces, the challenge is minimizing the difference between the two shapes in the spectral space. To align two surfaces using the Laplace–Beltrami spectral space, we aim to maximize the similarity between both eigenvalues and eigenvectors of the Laplace–Beltrami operator of the surfaces.

As a result of nonisometric deformation, the eigenvalues and eigenvectors of the shape dramatically change. On a compact closed manifold M with Riemannian metric g, we define the shape deformation as a time-variant positive scale function $\omega(t) : M \to R^+$ such that $g^{\omega}_{ij} = \omega g_{ij}$ and $d\sigma^{\omega} = \omega d\sigma$, where $\omega(t)$ is nonnegative and continuously differentiable.

To increase the similarity between eigenvectors, we try to minimize the distance of the eigenvectors on the feature points. Therefore we need a distance function in the embedding space. We employ the distance measurement proposed in [130]. To find the optimal scale function ω for two surfaces $(N, \omega g_1)$ and (M, g_2), the energy function is defined as

$$\begin{aligned} E(\omega, \Phi_1, \Phi_2) = &\int_N \left[d^{\Phi_2}_{\Phi_1}(x, M)\right]^2 d_N(x) \\ &+ \int_M \left[d^{\Phi_2}_{\Phi_1}(N, y)\right]^2 d_M(y), \end{aligned} \tag{6.4}$$

where $d^{\Phi_2}_{\Phi_1}(x, M)$ and $d^{\Phi_2}_{\Phi_1}(N, y)$ are defined as

$$\begin{aligned} d^{\Phi_2}_{\Phi_1}(x, M) &= \inf_{y \in M} \parallel I^{\Phi_1}_N(x) - I^{\Phi_2}_M(y) \parallel_2, \; x \in N, \\ d^{\Phi_2}_{\Phi_1}(N, y) &= \inf_{x \in N} \parallel I^{\Phi_1}_N(x) - I^{\Phi_2}_M(y) \parallel_2, \; y \in M, \end{aligned} \tag{6.5}$$

ω is the scale function applied on N, and Φ_1 and Φ_2 are the eigenvector bases for LB embeddings of $(N, \omega g_1)$ and (M, g_2). In this work, we focus on mapping one manifold to another and aim to find such a metric optimization. Therefore we assume that the manifold M is fixed and N changes using the scale function ω to minimize the distance between M and ωN on those feature points.

To increase the similarity of eigenvalues between two manifolds N and M, we employ the theorem proved in [43], that is, $\lambda_n(t)$ is piecewise analytic and, at any regular point, the t-derivative of $\lambda_n(t)$ is given by

$$\dot{\lambda_n} = -\lambda_n \int_M \dot{\omega} {f_n}^2 d\sigma. \tag{6.6}$$

This theorem shows that the spectrum is smooth and analytic to nonisometric local scale deformation.

6.3 Numerical optimization using spectral variation

In this section, we detail a discrete algorithm for the alignment of nonisometrically deformed shapes through the variation of eigenvalues and eigenvectors. Consider two closed manifolds N and M with eigenvalues λ_1 and λ_2 and eigenvector bases Φ_1 and Φ_2. These two manifolds are represented with discrete triangle meshes. We use their first k_1 nonzero eigenvalues and eigenvectors to align two surfaces. As we mentioned before, the deformation is not isometric;

thus the first k_1 eigenvalues and eigenvectors of these two surfaces are not the same. To align the first k_1 eigenvalues of N to those of M, a continuous scale diagonal matrix $\mathbf{\Omega(t)}$ is applied on N; $\mathbf{\Omega}$ is an m by m matrix, where m is the number of vertices on N. The element Ω_{ii} at the diagonal is a scale factor defined on each vertex on N and introduces a variation and alignment from N to M. It is a nonnegative continuously differentiable matrix.

To solve the numerical problem, we use the time interval of $t \in [0, 1]$ and divide it into K steps, which we will index as q. For each step of q, we solve an optimization equation to increase the similarity of eigenvalues and eigenvectors of $\mathbf{\Omega}N$ toward those of manifold M. At the beginning, $t = 0$, the eigenvectors and eigenvalues are Φ_1 and λ_1, and $\mathbf{\Omega}(0) = I$. When t reaches 1, the eigenvalues and eigenvectors will be λ_2 and Φ_2. To do that, we assume that the eigenvalues of N vary linearly toward those of M. This can be represented as

$$\lambda_n(t) = (1-t)\lambda_{1,n} + t\lambda_{2,n}, t \in [0, 1], \tag{6.7}$$

where n is the index of eigenvalues. Therefore, at any regular time of t, the derivative of λ is constant and can be calculated as

$$\dot{\lambda}_n(t) = \lambda_{2,n} - \lambda_{1,n}, t \in [0, 1]. \tag{6.8}$$

For mapping of eigenvectors, in each step, we minimize the distance function described in Eq. (6.4) between $\mathbf{\Omega}N$ and M. The following will explain the details how to calculate the optimization function to minimize the distance between eigenvalues and eigenvectors in each step.

6.3.1 Eigenvector optimization equation

To minimize the energy function in Eq. (6.4), we need to calculate the distance between two manifolds using Eq. (6.5). To do that, we compute the k_1 eigenvalues and eigenvectors for both manifolds using Eq. (6.2). One of the concerns about the calculating the eigenvectors is the sign ambiguity. This means that either f_n or $-f_n$ can be an eigenvector of a specific eigenvalue. For a target surface M, we fix the eigenvectors by picking random signs for Φ_2. Then we calculate the feature points as described before.

For the surface N, we start with $\mathbf{\Omega} = 1$, and at each step, we update the surface using the optimized $\mathbf{\Omega}$ to minimize the energy function E. At each step, we first calculate the k_1 eigenvalues and eigenvectors of the updated surface N using Eq. (6.2). Then we calculate the three sets of feature points using the eigenvectors as explained before. We need to find the corresponding feature sets on two surfaces for solving the sign ambiguity of eigenvectors. As shown in Fig. 6.1A, there is one nodal set for the eigenvector corresponding to the first nonzero eigenvalue. Therefore these sets are matched on two surfaces. For the other two nodal sets, we calculate the corresponding sets using their signs on the eigenvector corresponding to the first nonzero eigenvalue. We first calculate and determine the sign of this eigenvector using the histogram of the positive and negative eigenvector values for both surfaces. As we mentioned before, the second sets of the nodal nodes are parallel to the first set. Also, the sets of nodes are at the opposite sides of the first nodal set. Therefore each set of nodes has a different sign value on the first eigenvector. Knowing the sign of the first eigenvector, we can categorize and determine the corresponding sets of nodes for the second feature sets.

By detecting the corresponding feature sets at each step, we find the nearest feature points of surface N to the feature points of M that minimize the distance (6.5). To achieve this, we consider all combinations of signs for k_1 eigenvectors for surface N to minimize the distance equation. After finding the corresponding points and signs, the points are employed to generate the matrix $\mathbf{C}$ as the nearest feature points mapping from $I_N^{\Phi_1}$ to $I_M^{\Phi_2}$. This mapping can be presented as $Id(\mathbf{B}V_1) = \mathbf{C}V_2$, where V_1 and V_2 are the vectors that present the vertices of surfaces N and M, respectively. The matrix $\mathbf{B}$ is a diagonal matrix of the size of V_1 in which the diagonal elements for the feature points are 1 and 0 otherwise. The matrix $\mathbf{C}$ has the value 1 only for the feature points; therefore the projection relation Id can present a linear interpolation mapping from feature points of surface N to M. Using this mapping, we can write the energy function (6.4) in a discrete numerical form as

$$E = \sum_{n=1}^{k_1} \left(\frac{1}{\mathbf{S}(N)} \left(\frac{\mathbf{B} f_{1,n}}{\sqrt{\lambda_{1,n}}} - \frac{\mathbf{C} f_{2,n}}{\sqrt{\lambda_{2,n}}} \right)^T \mathbf{\Omega} \left(\frac{\mathbf{B} f_{1,n}}{\sqrt{\lambda_{1,n}}} - \frac{\mathbf{C} f_{2,n}}{\sqrt{\lambda_{2,n}}} \right) \right), \tag{6.9}$$

where $\mathbf{S}(N)$ is the surface area of N, and $\mathbf{\Omega}$ is the scale function. In this work, because we change the surface N toward the surface M and the surface M does not change at each step, the second part of Eq. (6.4) is zero, and only the first part is used to calculate the numerical equation. Considering that at each iteration the corresponding feature points are calculated and the eigenvalues do not change, the derivative of E with respect to time can be defined as follows:

$$E_f = \frac{\partial E}{\partial t} = \sum_{n=1}^{k_1} \left(\frac{1}{\mathbf{S}(N)} Ds_n^T \; \dot{\mathbf{\Omega}} \; Ds_n \right), \tag{6.10}$$

where $Ds_n = \left(\frac{\mathbf{B} f_{1,n}}{\sqrt{\lambda_{1,n}}} - \frac{\mathbf{C} f_{2,n}}{\sqrt{\lambda_{2,n}}} \right)$ and $\dot{\mathbf{\Omega}} = \frac{\partial \mathbf{\Omega}}{\partial t}$. Because $\mathbf{\Omega}$ is a diagonal matrix, we extract the diagonal elements as a vector $\mathbf{v}_{\mathbf{\Omega}}$, and Eq. (6.10) can be rewritten as

$$E_f = \sum_{n=1}^{k_1} \left(\frac{1}{\mathbf{S}(N)} ((Ds_n)^2)^T \mathbf{v}_{\dot{\mathbf{\Omega}}} \right). \tag{6.11}$$

Using this equation, we update the eigenvectors through the numerical optimization of the gradient of the energy function at each step.

6.3.2 Eigenvalue optimization equation

To increase the similarity of two eigenvalues of λ_1 and λ_2, we employ the method in [43]. Considering the scale function applied to the surface N, the weighted generalized spectral problem in Eq. (6.2) can be presented as follows:

$$\mathbf{W} f_n = \lambda_n \mathbf{\Omega S} f_n, \tag{6.12}$$

where λ_n and f_n are the corresponding nth solution. Using this equation, Eq. (6.6) can be transformed into the discrete form

$$\dot{\lambda_n} = -\lambda_n f_n^T \dot{\mathbf{\Omega}} \mathbf{S} f_n. \tag{6.13}$$

The function λ_n is piecewise analytic. Considering that we only apply $\mathbf{\Omega}$ to the surface N by combining equations (6.8) and (6.13), the derivative of each $\lambda_i(t)$ leads to the following equation:

$$-\lambda_{1,n}(t) f_{1,n}(t)^T \dot{\mathbf{\Omega}} \mathbf{S}_N f_{1,n}(t) = \lambda_{2,n} - \lambda_{1,n}, t \in [0, 1], \tag{6.14}$$

where $\mathbf{S}_N$ is a diagonal matrix where each element shows the Voronoi region for the corresponding vertex. Although the time derivative of $\mathbf{\Omega}$ can be calculated in Eq. (6.14), solving this equation is not straightforward. We need to transform the individual integration equation into a linear system. We achieve this by extracting the diagonals as vectors $\mathbf{v}_{\mathbf{\Omega}}$ and $\mathbf{v}_{\mathbf{S}_N}$ and then applying the Hadamard product to Eq. (6.14). Thus this equation can be rewritten in a linear form as follows:

$$(\mathbf{v}_{\mathbf{S}_N} \circ f_{1,n} \circ f_{1,n})^T \cdot \mathbf{v}_{\dot{\mathbf{\Omega}}} = \frac{\lambda_{1,n} - \lambda_{2,n}}{\lambda_{1,n}(t)}, t \in [0, 1]. \tag{6.15}$$

Note that as the first k_1 eigenvalues are going to be aligned, we can get k_1 independent equations, which lead to the linear system

$$\mathbf{a} \cdot \mathbf{v}_{\dot{\mathbf{\Omega}}} = \mathbf{b}, \tag{6.16}$$

where $\mathbf{a}$ is a row stack of $(\mathbf{v}_{\mathbf{S}_N} \circ f_{1,n} \circ f_{1,n})^T$ with k_1 rows, and $\mathbf{b}$ is the right-hand side of Eq. (6.15). Since we use the first k_1 eigenvalues for this work and practically k_1 is much less than the number of nodes in the mesh, the system is underdetermined and has no unique solution.

We solve this by assuming that the scale factors distributed on N are smooth. On the discrete triangle mesh N with the scale function vector $\mathbf{v}_{\mathbf{\Omega}}$, the smoothness energy of E is define as

$$E_\lambda =< \mathbf{v}_{\mathbf{\Omega}} + \mathbf{v}_{\dot{\mathbf{\Omega}}}, \mathbf{L}_N \cdot (\mathbf{v}_{\mathbf{\Omega}} + \mathbf{v}_{\dot{\mathbf{\Omega}}}) >_{\mathbf{S}}, \tag{6.17}$$

where $\mathbf{L}_N$ can be calculated using Eq. (6.1) for manifold N. Because the scale vector applies to the surface N only, this equation only applies to the surface N.

Assuming that $\mathbf{v}_{\Omega}$ is known at each time t, $\mathbf{v}_{\dot{\Omega}}$ is going to minimize the quadratic smooth energy. Then it leads to the equation

$$E_{\lambda} = \mathbf{v}_{\dot{\Omega}}^{T} \cdot \mathbf{W}_{N} \cdot \mathbf{v}_{\dot{\Omega}} + 2\mathbf{z}^{T} \cdot \mathbf{v}_{\dot{\Omega}}, \tag{6.18}$$

where $\mathbf{z} = \mathbf{W}_{N}\mathbf{v}_{\cdot\Omega}$. Through the combination of this energy function and the energy function calculated for eigenvectors, the distance between the eigenvalues and eigenvectors of two surfaces can be minimized.

6.3.3 Energy equation integration

To increase the similarity of eigenvalues and eigenvectors of two surfaces, we integrate the energy function calculated for both eigenvalues and eigenvectors to find a scale matrix that minimizes the total energy function as follows:

$$E_{T} = E_{\lambda} + E_{f}$$
$$= \mathbf{v}_{\dot{\Omega}}^{T} \cdot \mathbf{W}_{N} \cdot \mathbf{v}_{\dot{\Omega}} + 2\mathbf{z}^{T} \cdot \mathbf{v}_{\dot{\Omega}} + \sum_{n=1}^{k_1} \left(\frac{1}{\mathbf{S}(N)} ((Ds_n)^2)^T \mathbf{v}_{\dot{\Omega}} \right). \tag{6.19}$$

To preserve the physical availability, $\mathbf{v}_{\Omega}$ must be bounded, that is, the scale factor cannot be zero or negative, and it cannot be infinite either. We denote lower and upper bounds by $\mathbf{h}_l, \mathbf{h}_u > \mathbf{0}$, where $\mathbf{h}_l$ and $\mathbf{h}_u$ are n-dimensional constant vectors, and $\mathbf{v}_{\dot{\Omega}}$ must satisfy

$$\mathbf{h}_l \leq \mathbf{v}_{\Omega} + \mathbf{v}_{\dot{\Omega}} \leq \mathbf{h}_u. \tag{6.20}$$

This inequality bound can be written in a matrix form:

$$\mathbf{G} \cdot \mathbf{v}_{\dot{\Omega}} \leq \mathbf{h}, \tag{6.21}$$

where $\mathbf{G}$ is a stack of identity matrices,

$$\mathbf{G}_{2n\times n} = \begin{pmatrix} -\mathbf{I}_{n\times n} \\ \mathbf{I}_{n\times n} \end{pmatrix}, \tag{6.22}$$

and $\mathbf{h}$ is a $2n$-dimensional vector,

$$\mathbf{h}_{2n\times 1} = \begin{pmatrix} \mathbf{v}_{\Omega} - \mathbf{h}_l \\ \mathbf{h}_u - \mathbf{v}_{\Omega} \end{pmatrix}. \tag{6.23}$$

The linear system (6.16), energy function (6.19), and constant bound (6.21) form a quadratic programming problem at each time t. Assume that the eigenvalues and eigenvectors are known at each time t and the derivative of the scale matrix

$\dot{\mathbf{\Omega}}$ is the solution of this problem. The result $\dot{\mathbf{\Omega}}(q)$ for each iteration can be used to calculate $\mathbf{\Omega}(q+1)$ as follows:

$$\mathbf{\Omega}(q+1) = \mathbf{\Omega}(q) + \frac{1}{K-q}\dot{\mathbf{\Omega}}(q). \tag{6.24}$$

After K steps, the desired $\mathbf{\Omega}(K)$ will be achieved, and two manifolds are aligned. The summary of the algorithm can be found in Algorithm 2. As shown, after K steps the surface N will be aligned to surface M, and the correspondence can be computed using the aligned eigenvectors.

Algorithm 2 Spectrum alignment.

Input: Closed 2D manifolds N and M represented by triangle meshes and constant k_1;

Output: Diagonal weight matrix $\mathbf{\Omega}(q)$ on N, aligning the first k_1 nonzero eigenvalues and corresponding eigenvectors of feature points from N to M;

1: Initialize $\mathbf{\Omega}(0) \leftarrow \mathbf{I}$, calculate the matrices $\mathbf{W}_N$ and $\mathbf{S}_N$, and $\lambda_{2,n}, f_{2,n}, \lambda_{1,n}$, and $f_{1,n}$ for $n = 1, 2, \ldots, k_1$;

2: Compute the feature points for surfaces M.

3: **while** $q < K$ **do**

a: Calculate $\lambda_{1,n}(q), f_{1,n}(q)$ for $n = 1, 2, \ldots, k_1$ using Eq. (6.2) with $\mathbf{\Omega}(q)$;

b: Calculate the feature points for surface $\mathbf{\Omega}(q)N$ using $f_{1,n}(q)$; solve the eigenvector sign ambiguity and find the corresponding feature points between surfaces $\mathbf{\Omega}(q)N$ and M;

c: Construct the quadratic programming problem using Eqs. (6.16), (6.19), and (6.21);

d: Solve the quadratic programming problem to get $\dot{\mathbf{\Omega}}(q)$ and calculate $\mathbf{\Omega}(q+1)$;

e: $q \leftarrow q+1$;

4: **end while**

5: The correspondence of the surfaces N and M can be computed using the aligned eigenvectors.

6.4 Experiments and applications

The proposed algorithm and system are implemented using Python and C++ on a 64-bit Linux platform. For visualization purposes, we employ MATLAB® and VTK library in Python. The experiments are conducted on a computer with an Intel Core i7-3770 3.4 GHz CPU and 8 GB RAM. We apply our algorithm to 2D manifolds represented by triangle meshes. We employ the approach in [157, 151] to generate the uniform meshes. The number of vertices in those meshes is about 3000 for most of the experiment data. Besides the vertex number, there are two constants, that is, K iteration and the first k_1 nonzero eigenvalues and eigenvectors to be aligned. For our experiments we choose $K = 10$. This number is sufficient to generate accurate results. We use different k_1 for different experiments. Depending on the resolution that we need in our experiments, the number k_1 may vary. The average computational time for 3000 nodes with $k_1 = 8$ and $K = 10$, is around 12 seconds.

6.4.1 Experiments on synthetic data

6.4.1.1 Our results

To evaluate our method, we manually make some nonisometric deformations on the surface of the shape, and then we register the initial shape to the deformed one. In these experiments, we use a Stanford bunny model and make a nonisometric deformation on the surface and then generate uniform triangle meshes on both surfaces. We employ the first 10 nonzero eigenvalues and the corresponding eigenvectors to do the alignments. The processing time for $k_1 = 10$, $K = 10$, and 3000 mesh vertices is about 43 seconds. Note that no correspondence information is used in the experiments.

In the first experiment, we manually generate a bump on the back of a bunny and align the original surface to the deformed one. Fig. 6.2 shows the original

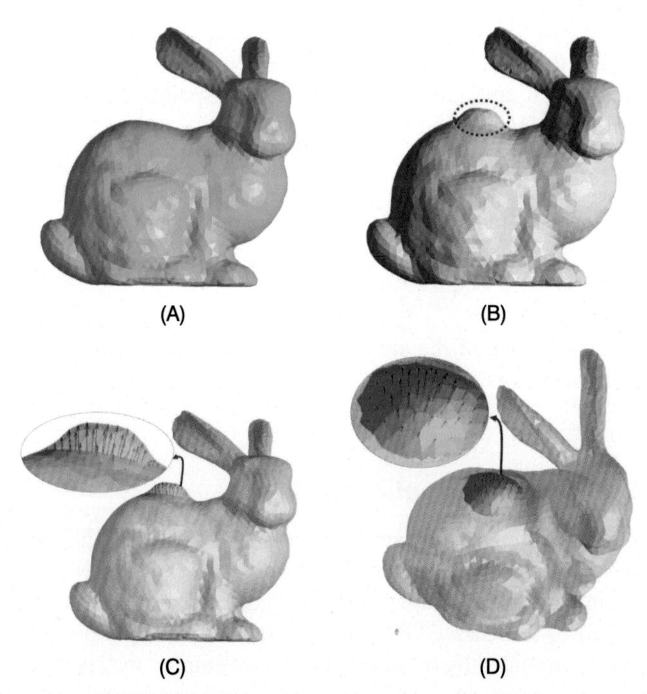

FIGURE 6.2 The result of mapping the original 3D object to the synthetic one. (A) shows the original object. (B) is obtained by generating a bump on the original surface. (C) and (D) show the results of point-to-point mapping the original surface (cyan) to the target one (yellow) from different angles. (Image taken from [37]. ©2019 IEEE. Included here by permission.)

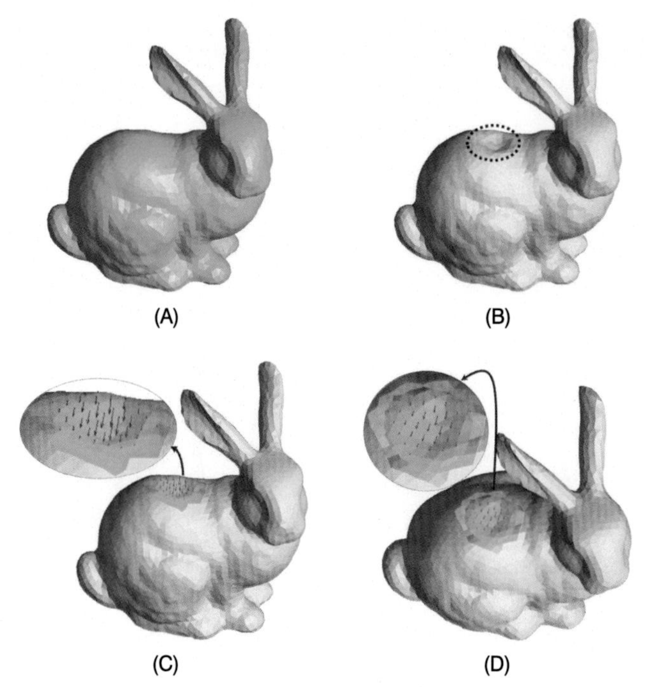

FIGURE 6.3 The result of mapping the original 3D object to the synthetic one. (A) shows the original object. (B) is obtained by generating an indentation on the original surface. (C) and (D) show the results of point-to-point mapping the original surface (cyan) to the target one (yellow) from different angles. (Image taken from [37]. ©2019 IEEE. Included here by permission.)

surface in cyan and target surface in yellow. The location of the bump is marked by a red circle. Figs. 6.2C and 6.2D present the results of point-to-point mapping of the surfaces from different angles. The original and targeted shapes are overlaid, and the arrows in the bump area show the deformation of each vertex from the original to the targeted surface.

In the second experiment, we manually create an indentation on the surface of a bunny and align the original surface to the dent one. Figs. 6.3A and 6.3B show the original and surface results of creating the nonisometric dent on the surface, respectively. Figs. 6.3C and 6.3D present the result of point-to-point mapping of surfaces using our method. The original and targeted shapes are overlaid, and the arrows in the bump area show the deformation of each vertex from the original to the targeted surface. These results confirm that our method can accurately detect and localize the nonisometric deformation and find the corresponding points.

For more complex and challenging deformations, we use a hammer model and create 2000 uniform mesh vertices on the surface. Then we create different nonisometric deformations on the surface and align the original surface to the deformed one. Fig. 6.4 shows the original hammer in yellow color and target

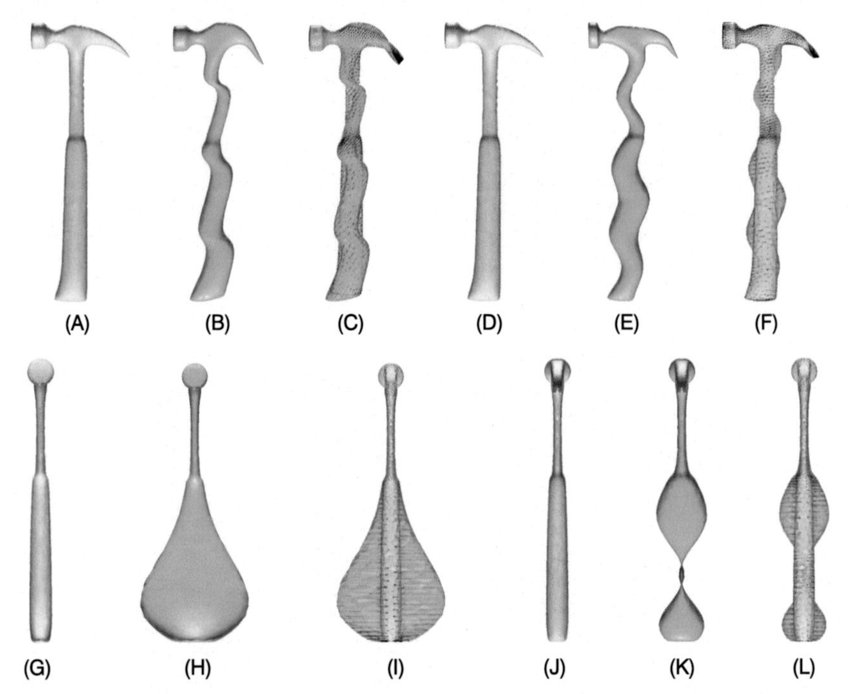

FIGURE 6.4 The result of mapping the original 3D hammers to the synthetic deformed ones. (A), (D), (G), and (J) show the original hammers. (B), (E), (H), and (K) are obtained by generating nonisometric deformation on the original surface. (C), (F), (I), and (L) show the results of point-to-point mapping the original surface (yellow) to the target one (cyan). (Image taken from [37]. ©2019 IEEE. Included here by permission.)

deformed hammers in cyan color. Figs. 6.4C, 6.4F, 6.4I, and 6.4L show the results of aligning the original surface to the deformed ones using our method. The point-to-point alignments are demonstrated using the arrows that connect the corresponding points on the original and deformed surfaces. The original and target surfaces are overlaid in these figures. These results show that our method can accurately detect simple and complex nonisometric deformations on the surface.

6.4.1.2 *Comparison to a spatial-based method*

Too further demonstrate the capabilities of our method, we compare the results of our algorithm with those from nonrigid Iterative Closest Point (ICP) algorithm. ICP is introduced by Besl and Mckay [8] and is one of the popular approaches in spatial registration-based methods. In this approach, the initial transformation for global matching is first estimated, and then the closest points are found by minimizing the distance between two shapes. Therefore, using this method, we first register the original surface to the target one rigidly, and then the corresponding points between the rigidly registered original and target

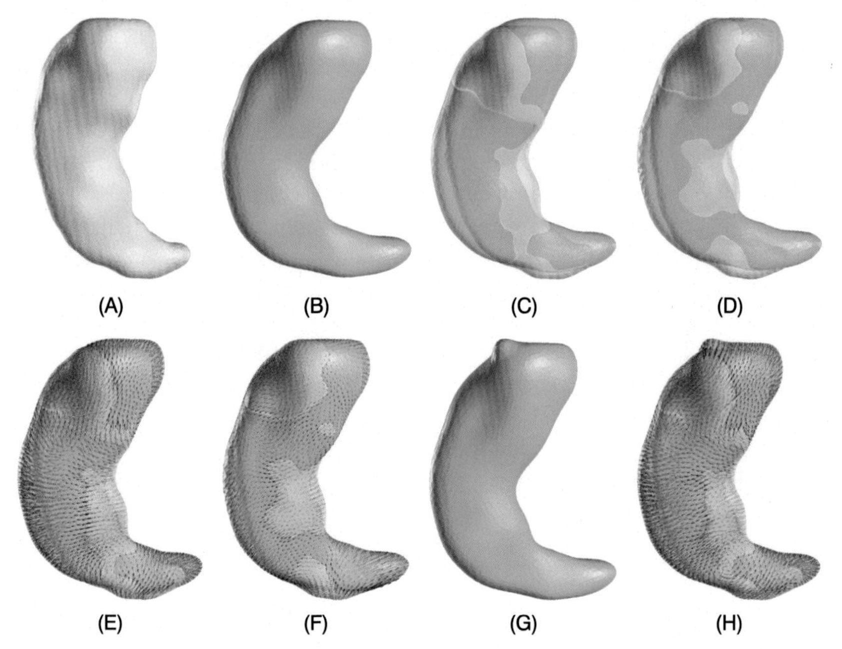

FIGURE 6.5 Comparison of our method with ICP method using synthetic data. (A) presents the original surface, and (B) is obtained by bending and stretching the shape from the upper and lower ends. (C) shows the result of mapping these two shapes. (D) shows the result of ICP rigid registration result. The ICP method registers the shape from one side, and therefore this method cannot generate accurate result for bending deformation. (E) and (F) present the results of point-to-point mapping from the original surface to the deformed one using our method and ICP method, respectively. Because the result of nonrigid ICP depends on rigid ICP, the result is not accurate. Our method can detect the deformation accurately. To show that our method can simultaneously handle both global and local deformations, we make a bump on the deformed surface as presented in (G). The result of mapping (A) to (G) is presented in (H). (Image taken from [37]. ©2019 IEEE. Included here by permission.)

shapes are calculated. For our method, we employ the first 12 nonzero eigenvalues and the corresponding eigenvectors for alignment purposes. The processing time for 3000 vertices, $k_1 = 12$, and 10 iterations is 118 seconds.

To compare two methods, we use a template hippocampus and synthetically deformed the shape by bending and stretching the shape from upper and lower sides. The original and deformed surfaces are shown in Figs. 6.5A and 6.5B, respectively, which exhibit global variation. Fig. 6.5C presents the overlay of the original and deformed shapes. As can be seen, the top and bottom parts of the surface are stretched, and the shape is bent in the middle part. Fig. 6.5D shows the result of performing ICP rigid registration on the original shape to map it to the deformed shape. Comparing Figs. 6.5C and 6.5D, we can notice that the rigid ICP does not match the shapes correctly, especially in the top and bottom regions of the shapes. Figs. 6.5E and 6.5F present the results of our method and nonrigid ICP, and the arrows show the displacement of each vertex on the surface. Because the ICP method fails in the rigid registration stage, the corresponding points calculated using nonrigid ICP do not reflect the accurate deformation, especially in the top and bottom regions. On the other hand, because our method does not require prerigid registration to find the corresponding points, this variation can be accurately captured by our registration and mapping method. These results justify the advantage of our method over the rigid and nonrigid ICP methods. To demonstrate that our method can simultaneously handle both global and local deformations, we create a bump on the shape of Fig. 6.5B as shown in Fig. 6.5G. Then we align the surface in Fig. 6.5A to Fig. 6.5G using our method. Both global and local deformations (bump) can be captured by our method as shown by arrows in Fig. 6.5H. Therefore these results confirm that our method can detect and localize the nonisometric deformation and find the correspondence and their displacements resulted from both global and local deformations.

6.4.1.3 *Comparison to a spectral-based method*

To compare our method with a similar spectral-based technique, we employ an approach suggested by Shi et al. [130]. They proposed a method based on aligning the eigenvectors of two surfaces via optimization of a conformal metric on the surfaces. They employed the eigenvectors for all the points of the surfaces, and therefore the computation is very expensive.

In this experiment, we employ a template hippocampus and create 1500 uniform vertices on the surface. Then, similarly to the previous section, we synthetically deform the shape by bending and stretching the shape from upper and lower sides. The bending and stretches are larger in this experiment than in the previous one. Figs. 6.6A and 6.6B show the original and synthetically deformed surfaces, respectively. Figs. 6.6D and 6.6C show the results of the alignment using Shi et al.'s and our methods. As can be seen, the bottom tip of the sur-

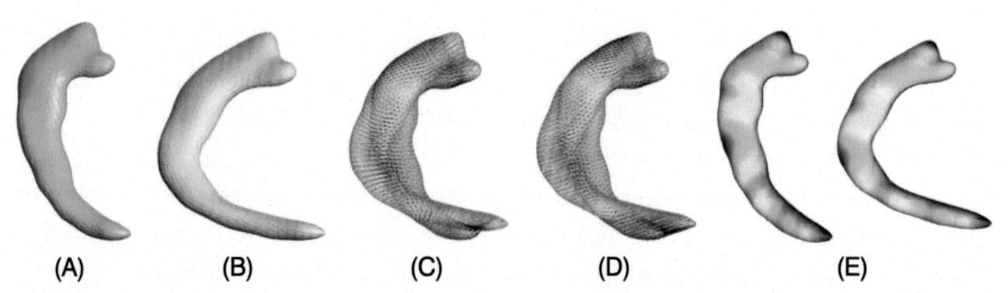

FIGURE 6.6 Comparison of our method with Shi et al.'s method [130] using synthetic data. (A) presents the original surface, and (B) is obtained by bending and stretching the shape from the upper and lower ends. (C) shows the result of Shi et al.'s approach. The lower tip of the shape is not aligned correctly using Shi et al.'s method, whereas our method can align all the points accurately. (D) shows the result of mapping these two shapes using our method. (E) presents the color map visualization of the corresponding points of the shapes using our method. (Image taken from [37]. ©2019 IEEE. Included here by permission.)

TABLE 6.1 Comparison among our method, Shi et al.'s method, and nonrigid ICP method. (Table taken from [37]. ©2019 IEEE. Included here by permission.)

Capabilities	Our method	Shi et al.'s method	Nonrigid ICP
Nonrigid registration	✓	✓	
Local deformation	✓	✓	✓
Average accuracy: A	91.4%	85.8%	75.6%
Computation	< 120 s	> 20 m	> 60 s

face cannot be aligned correctly using Shi et al.'s method, but our method can accurately align all the points.

To quantitatively evaluate the capabilities of these methods in localizing the point-to-point correspondence, we use the following metric:

$$A = 1 - \frac{\sum_{i=1}^{m} |d_i - d_i^O|}{\sum_{i=1}^{m} |d_i^O|}, \tag{6.25}$$

where d_i is the distance between corresponding points calculated using either methods, d_i^O is the known ground truth distance, m is the number of all nodes, and i is the index of nodes. The experiments demonstrate that the average outcome for our method is 91.4%, whereas it is 85.8% for Shi et al.'s method. This number is 75.6% for ICP method. Therefore our method can find the point-to-point correspondence better than the other two methods for complex deformation. In this experiment, we use 10 iterations and 12 nonzero eigenvalues and eigenvectors to do the alignments. For 1500 mesh vertices, our method takes 67 seconds, whereas Shi et al.'s approach takes 94 minutes. In [130], they mentioned that their execution time for 1000 triangle faces with approximately 500 vertices is 20 minutes.

Table 6.1 demonstrates the comparison between our method, Shi et al.'s method, and nonrigid ICP method. As mentioned before, the nonrigid ICP method requires rigid registration before aligning two surfaces, whereas Shi et al.'s and our methods do not have this requirement. All the methods can localize the deformation of the surface, but our method has the best average accuracy based on metric defined in Eq. (6.25). The computational time for our method is considerably less than Shi et al.'s method and similar to nonrigid ICP. Therefore our method has the best features to align two surfaces.

6.4.2 Applications on real patient imaging data

6.4.2.1 *Alzheimer data*

Alzheimer disease (AD) is a brain misfunctionality caused by the loss of neurons and neural volume. Hippocampus is vulnerable to damage in the early stage of Alzheimer. Volumetric longitudinal studies using MR images show hippocampal atrophy during time in comparison to healthy cases. In this study,

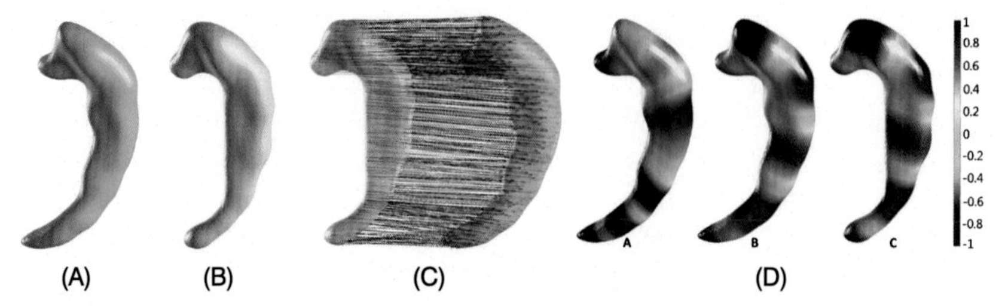

FIGURE 6.7 (A) shows the baseline hippocampus, and (B) shows the hippocampus for the same subject after one year. (C) presents the result of mapping the baseline hippocampus to the one after one year. (D) presents the 6th eigenvector before and after mapping. Column A shows the eigenvector before alignment, and column B shows the eigenvector after alignment. Column C shows the 6th eigenvector for the target surface. (Image taken from [37]. ©2019 IEEE. Included here by permission.)

TABLE 6.2 The result of aligning eigenvalues from the baseline hippocampus to one after one year (target) using the same case as in Fig. 6.7. (Table taken from [37]. ©2019 IEEE. Included here by permission.)

Manifold	λ_i/λ_1, $i \in [2, 10]$
Baseline	3.40, 7.39, 10.66, 15.36, 17.04, 21.50, 23.13, 24.94, 29.77
Target	4.17, 8.65, 11.42, 16.23, 18.97, 23.18, 26.19, 30.53, 32.38
Aligned	4.17, 8.65, 11.42, 16.23, 18.96, 23.17, 26.19, 30.49, 32.39

we show the point-to-point deformation for the Alzheimer case. We employ 10 AD and 10 healthy cases, which have longitudinal study for one year to track and compare the deformation of hippocampi. The cases are downloaded from the hippocampal study from the Alzheimer's Disease Neuroimaging Initiative (ADNI) database (adni.loni.usc.edu) and are segmented using FreeSurfer software. Then the 3D objects and meshes with 3000 vertices are generated. In this study, we use the first 10 eigenvectors to align the surfaces. Figs. 6.7A and 6.7B show a sample of AD case for the baseline and after one year. The deformation in the tail part can be detected visually. The deformation mapping is shown by using the blue arrows in Fig. 6.7C. We need to mention that the deformation mapping is down-sampled by five in order to better visualize the results. As can be seen, our method can accurately detect the deformation. Fig. 6.7D shows the 6th eigenvector before and after alignment. Columns A and B show the eigenvector for the baseline surface before and after alignment, respectively. Column C shows the targeted surface eigenvector. Note that our method can match and align the eigenvectors. The color map shows the scaled value of the eigenvector.

To show the variation of eigenvalues of the manifolds before and after alignment, we list the 2nd to 10th nonzero eigenvalues of baseline hippocampus (before and after mapping) and hippocampus after one year in Table 6.2. The eigenvalues are normalized by the first nonzero one to remove the scale factor.

We can see that after applying the spectrum alignment algorithm, the eigenvalues of the source manifold have changed to well align with the target ones.

6.4.2.2 Cardiac data

We use the Sunnybrook Cardiac Data [113] for this experiment. The data is acquired from the 3D left ventricle during cardiac cycles from end diastolic to end systolic and then back to end diastolic cycle. For this study, we use five cases for each category: heart failure with infarction, heart failure without infarction, left ventricle (LV) hypertrophy, and healthy. Meshes with 5000 vertices are generated. 10 nonzero eigenvalues and the corresponding eigenvectors are used for alignment. We map the surface for the end diastolic (first time point) to all other 19 time points. Then the mean averages of the displacements for all the nodes are calculated and used to generate a plot of surface displacement for 19 different time points. Fig. 6.8A shows the result of mapping these surfaces using our method and the mean average plot for a sample case. The first shape close to the origin of the axes is the diastolic shape, and the rest of the shapes are the target ones. As can be seen, the results follow the heart beating pattern. Fig. 6.8B shows the overlay of the first time point surface (cyan) to the 7th one (yellow), which has the most mean displacement according to the average plot. The results of mapping these two surfaces are also presented in this figure by using arrows, which connect the corresponding points and show the point-to-point deformation mapping. We can see that our method can detect the contraction and the tangential turning deformation of the heart.

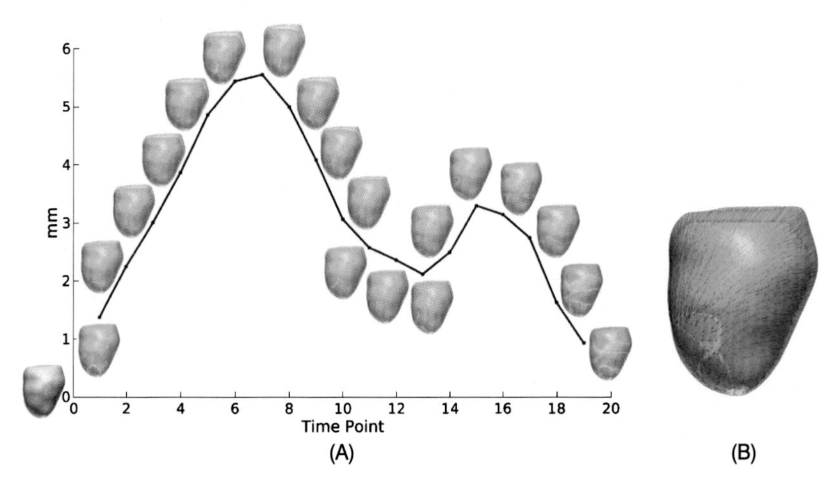

FIGURE 6.8 (A) The result of mapping the first time point shape to all other shapes and then computing the mean average of the distance. The yellow surface is the left ventricle in diastolic state. The other shapes shows the contracted left ventricle toward systolic state overlaid on the yellow surface. (B) The result of mapping the first time point surface (cyan) to the 7th one (yellow), which has the most deformation according to the plot in (A). (Image taken from [37]. ©2019 IEEE. Included here by permission.)

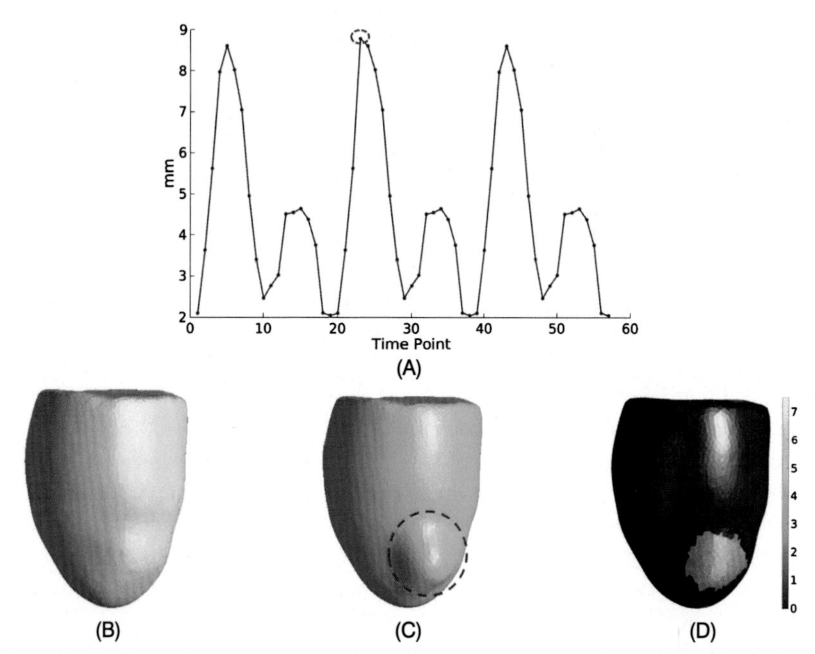

FIGURE 6.9 The longitudinal study for a subject using our method. By studying more than one cycle the abnormal beat can be detected using our method, and the abnormal area of the left ventricle can be identified. (A) shows 3 heart beat cycles, and the abnormal time point is marked by a red circle. We create this abnormality by creating a bump on a heart surface. (B) shows the original heart. (C) shows the deformed surface. The deformed area is marked by a red circle. (D) shows the displacement color map generated by our method to detect the abnormal region. (Image taken from [37]. ©2019 IEEE. Included here by permission.)

This method can be used in different applications in cardiac study. The first application is longitudinal study of a subject. In this study the 3D images of the subject are generated for more than one heart beating full cycles. Using these models, the deformation plot for mean average of displacement can be generated. Physicians usually look at more than one cycles to exam the abnormality of the heart. By using our method the abnormality can be detected via the mean average plot and then using the point-to-point deformation mapping to find out the specific abnormal time points. Therefore the abnormal area of the ventricle may be detected. Fig. 6.9A shows three heart beating cycles and the one time point, which is marked by a red circle in the 2nd cycle showing abnormality according to its location. As can be seen, the abnormality can be detected using our method. We detect this abnormality by creating a bump (Fig. 6.9C) on a heart surface (Fig. 6.9B). Then our method can generate the displacement color map, and the abnormal region can be also located (Fig. 6.9D).

The second application is the cross-subject analysis through temporal alignment. The same time point of a heart beat cycle for two different subjects can be identified. For instance, in one subject the largest heart contraction may happen

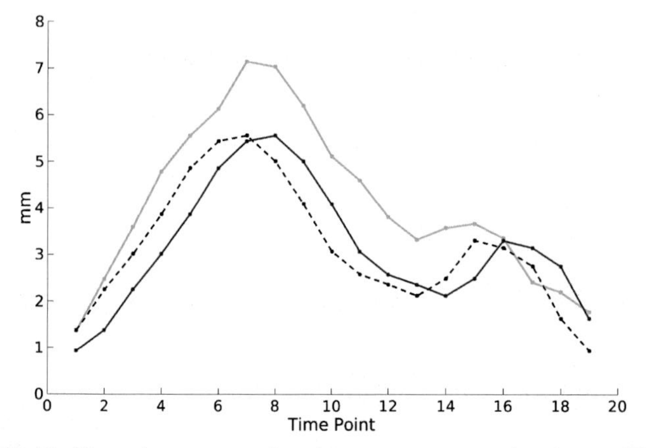

FIGURE 6.10 The blue and green curves show the mean average mappings for two different cases. The black dashed curve shows the blue curve after alignment. (Image taken from [37]. ©2019 IEEE. Included here by permission.)

at the 7th time point, but in another subject the largest heart contraction may be detected at the 8th time point. Using our method, different time points in different subjects can be aligned according to the heart beating cycle through temporal alignment. Fig. 6.10 shows this procedure. Using this alignment, we can align different healthy cases and then use the aligned mean average mapping to generate a standard mean average mapping. This mapping can be used to compare with normal and patient cases in order to categorize the abnormal cases from the normal ones by using the difference indicator between this mapping and a target one. In Fig. 6.11 the red plot shows the standard mean average plot generated by calculating the average of mean average mappings from temporally aligned five healthy cases. Fig. 6.11A shows a healthy case in a blue curve, as compared with the standard mean average mapping. Fig. 6.11B shows a diseased case in a green curve when temporally aligned to the standard mean average mapping, and Fig. 6.11C shows an LV hypertrophy case in a black curve. Yellow area shows the difference between two curves. As can be seen, the difference area in healthy case is much less than that in the patient cases. Using difference area, we can accurately categorize healthy cases from patient cases. Figs. 6.11D, 6.11E, and 6.11F show the displacement mapping of the left ventricle from the diastolic to systolic state for normal, diseased, and hypertrophy cases showed in Figs. 6.11A, 6.11B, and 6.11C, respectively.

6.4.3 Application on hand data

To demonstrate that our method can be used on nonmedical applications as well, we use data of different hand gestures and create 3000 uniform vertices on the surfaces. Figs. 6.12A, 6.12B, and 6.12C show different gestures of the hand. We employ the shape in Fig. 6.12A as the original surface and then align it to the

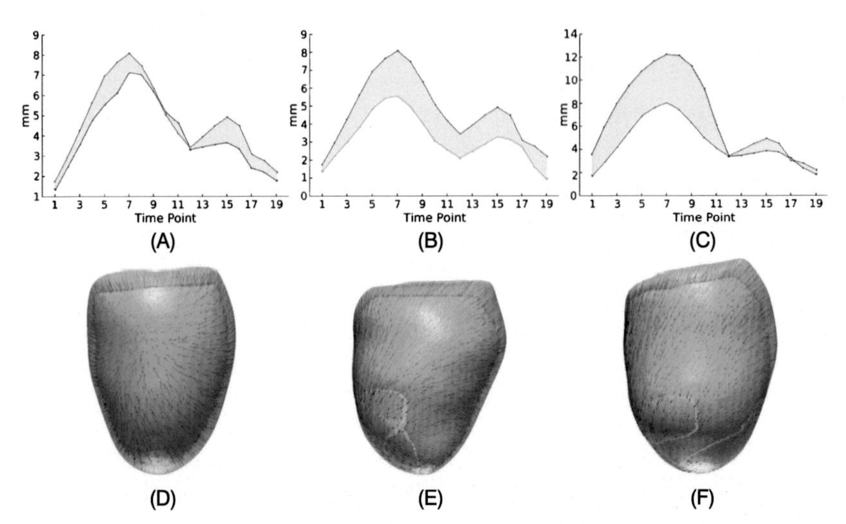

FIGURE 6.11 The result of comparing the mean average of five temporally aligned healthy cases to a healthy and two patient cases. In all plots the red curves show the average plot generated by calculating the average of five aligned healthy cases' mean average mappings. (A) The blue curve shows a healthy case. (B) The green curve shows a diseased heart case. (C) The black curve shows a hypertrophy case. We can use the difference between two curves (yellow area) to accurately distinguish the healthy from patient cases. (D), (E), and (F) show the displacement mapping of the left ventricle from the diastolic to systolic state for normal, diseased, and hypertrophy cases showed in (A), (B), and (C), respectively. (Image taken from [37]. ©2019 IEEE. Included here by permission.)

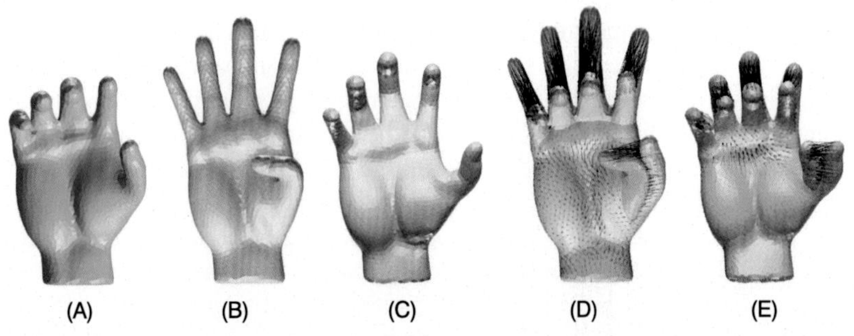

FIGURE 6.12 The results of aligning different hand gestures. (A) shows the original surface. (B), and (C) show the target surfaces. (D) and (E) show the results of aligning original surface to the surfaces demonstrated in (B) and (C), respectively. The results show that our method can correctly detect the point-to-point correspondence between two surfaces. (Image taken from [37]. ©2019 IEEE. Included here by permission.)

shapes in Figs. 6.12B and 6.12C. Figs. 6.12D and 6.12E show the results of this alignment using our method. The original and target surfaces are overlaid, and the point-to-point alignment is shown using the arrows that connect the corresponding points. As can be seen, different parts of the hand are correctly aligned, and our method can accurately detect the point-to-point correspondence.

6.5 Summary

In this chapter, we have introduced a new method based on the shape spectrum to find the point-to-point correspondence between two surfaces undergoing global and local deformations. We employ both eigenvalues and eigenvectors of the surfaces in the alignment. Our method can localize the nonisometric deformation of the surface and find the displacement mapping for all the vertices. Because we use certain feature points instead of all the vertices to align the eigenvectors, our method is considerably more efficient than the existing methods. We have applied our method to both synthetic and real data, and the results confirm the advantage and accuracy of our method. We have also compared our method with nonrigid ICP method and a similar spectrum-based method. The results show that our method has the best accuracy. In terms of computational time, our method is considerably faster than the previous spectrum-based method and similar performance as nonrigid ICP method.

Note that, for searching for proper eigenvectors, in addition to sign ambiguity, there are some cases that the order of the eigenvectors switches. In addition, by using large number of eigenvectors numerically it is possible for near multiplicities of eigenvectors to cause the eigenspaces to split in different directions [130]. In such cases the eigenvectors matching becomes difficult. In our algorithm, we mainly focus on resolving the sign ambiguities. The order switching of eigenvectors and detection of high-dimensional multiple eigenvectors will be our future research work.

Chapter 7

Deep learning of spectral geometry

Contents

In the previous chapters, all the methods are model-driven techniques for either spatial domain or spectral domain. The goal of this chapter is to describe the main mathematical ideas behind geometric deep learning and to provide development details for several applications in shape analysis and synthesis, especially in spectral geometry. The deep learning-based geometry is a data-driven technique, and we consider it as data geometry. The materials in this chapter are primarily based on recently published work in 3D geometric deep learning. With these materials, we gather and provide a clear picture of the key concepts and techniques, and illustrate the related applications. We also aim to provide practical implementation details for the methods presented in these works.

Spectral Geometry of Shapes. https://doi.org/10.1016/B978-0-12-813842-7.00015-2

Point cloud

Surface mesh

Voxelization

Projection view

FIGURE 7.1 Different 3D shape representations for a Bunny model.

7.1 Domain generation for learning of geometry

Deep learning methods have been penetrated into various research areas such as computer vision, natural language processing, audio processing, and so on. Specifically, they have contributed to a tremendous amount of applications and solved a bunch of tough tasks in computer vision due to the introduction of convolutional neural networks (CNNs), which was first proposed by LeCun et al. [75]. After that, great progress has been made on the improvements and extensions of CNNs. Nowadays, CNNs have played a crucial and extremely competitive role in 2D image understanding tasks [69]. For instances, it can save a lot of labor work and increase accuracy in image classification, object detection, and semantic segmentation. As we know, deep learning is a data-driven research method, and thus an adequate high-quality 2D dataset acquired from multiple resources is one of the most important factors for the current achievements.

Meanwhile, with the rise of the low-cost 3D acquisition devices (e.g., the Microsoft Kinect) and 3D modeling tools, increasing amount of 3D models from user-end are available. Many researchers have realized the importance and urgency to explore more functional and advanced methods to analyze and understand such 3D models so as to satisfy the large demand from the computer graphics and vision fields. The deep learning methods have been proved to be quite successful in 2D image area and now are a good way to try on 3D models. However, we must admit that there are a bunch of challenges in deep learning for 3D models. For example, 3D models have more complicated and variant representations, such as volumetric representations, multiple projected view image representations, unstructured point cloud, and manifold representations as shown in Fig. 7.1. How to classify, segment, and retrieve these 3D shapes and models has become very popular in recent years.

Many 3D deep learning architectures have been proposed targeting at different tasks, which can be categorized according to different criteria. For instance, we can divide the methods into three types based on how the 3D models are represented. The first type of deep learning architectures applies an image-based method to 3D models by either converting them into volumetric 3D images or multiview 2D images; the second type of deep learning architectures is designed to deal with 3D point clouds; the third type is suitable for (mesh) manifold representations that contain connectivity information. Meanwhile, we can divide 3D deep learning architectures based on the spatial and spectral domains as well. All these methods have been proved to be effective in terms of their specific tasks. Most of them are established on the CNNs baseline, and there are also other specially designed deep learning frameworks. We will introduce them in the following sections. Note that the challenges are also everywhere considering the 3D nature, the quality, and the amount of the dataset.

7.1.1 Challenges in geometric deep learning

Dealing with speeches, images, or videos on 1D, 2D, and 3D Euclidean domains, respectively, has been the main focus of research in deep learning for the past decades. However, in the recent years, more and more fields have to deal with data on non-Euclidean geometric domains, such as 3D surface models. For instance, in computer graphics and vision, 3D shapes are modeled as Riemannian manifolds (surfaces) endowed with properties such as curvature, color texture, or motion field (e.g., dynamic meshes). Even more complex examples include networks of operators, such as functional correspondences [107] or difference operators [122] in a collection of 3D shapes, or orientations of overlapping cameras in multiview vision (structure from motion) problems.

On the one hand, there is a strong need: the complexity of geometric data and the availability of large datasets, such as large-scale indoor and outdoor 3D scenes, social networks, and so on, make it tempting and very desirable to resort to machine learning techniques. On the other hand, there is a major challenge: the non-Euclidean nature of 3D geometric data implies that there are no such familiar properties as global parameterization, common system of coordinates, vector space structure, or shift-invariance. Consequently, basic operations such as linear combination or convolution, which are taken for granted in the Euclidean case, are not well defined on non-Euclidean domains. Currently, this happens to be a major obstacle to successfully use deep learning methods such as convolutional or recurrent neural networks on non-Euclidean geometric data. As a result, although there is a wide breakthrough in deep learning methods for natural language processing, image processing, and computer vision, it has not yet come to fields such as computer graphics, geometric modeling, and so on.

7.1.2 Basics of convolutional neural networks

Firstly, in this subsection we give a brief introduction of convolutional neural networks (CNNs). In machine learning, CNN is a class of deep, feed-forward artificial neural networks (ANNs) that has widely been applied to analyzing visual imagery. Just like the common ANNs, it consists of neurons that have learnable weights and biases. Each neuron receives the inputs and operates the

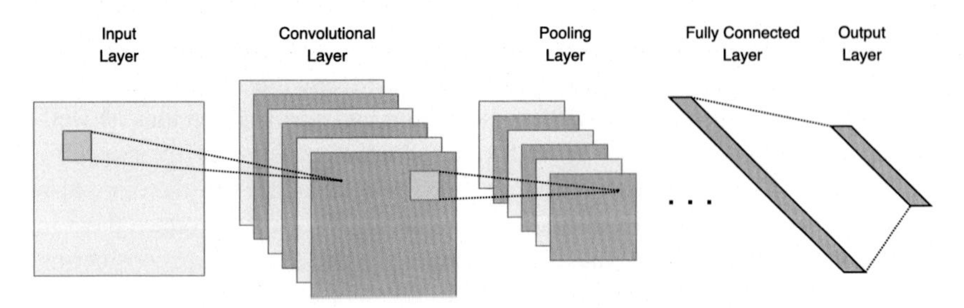

FIGURE 7.2 The basic structure of CNNs.

dot product accompanied with activation function to make the output nonlinear. CNN also has a loss function at the end of the architecture, and the training strategies are similar to other ANNs. There are three main types of layers in CNNs: convolutional layer, pooling layer, and fully connected layer. Fig. 7.2 shows the basic components in a CNN.

Convolutional Layer: Convolutional layer is the core module in CNNs and responsible for most of the computations. Each convolutional layer consists of several different filters. Every filter is a group of neurons that detect a particular feature for the layer inputs, which operates like a sliding rectangular window across the width and height of the input and computes the dot products between the filter entries and the values of input pixels covered by the filter. A 2D activation map is produced accordingly and is the response of the filter at every spatial position. During the convolution procedure, the step that the sliding window takes, also called stride, decides the size of the activation map. What kind of feature is extracted is determined by the weights associated with the filter. The final output of one convolutional layer is a stack of the activation maps produced by filters containing different weights. The number of the activation maps is called the depth or channel of the convolutional layer. It determines the number of features this convolutional layer is able to extract.

Pooling Layer: It is common to periodically add a pooling layer between two successive convolutional layers. It serves as a role to progressively reduce the redundant spacial size of the representations received from the previous layer. As a result, not only the amount of the network parameters and the computation can be reduced, but also the potential overfitting problem of the CNN model can be alleviated. The pooling layer operates independently on every depth slice from the previous layer and downsamples it spatially. There are two common types of pooling, max pooling and average pooling. Max pooling is most frequently applied in CNN; for instance, a 2×2 filter operates on the depth slice with a stride of 2 by extracting every maximum of the local four pixels covered by the pooling filter. Since the filter slides across the whole single depth slice with no overlap, the pooling operation can reduce the input spatial size by 75%. Besides that, there are some other pooling functions, such as L_2-norm pooling and so on.

Fully Connected Layer: In a fully connected layer the neurons have full connections to all activations in the previous layer, which means that it no longer contains any spatial information in this layer operation. Fundamentally speaking, their activations are computed with a matrix multiplication followed by a bias offset. The functionality of a fully connected layer is, in some terms, organizing and reshaping the result from successive convolutional layers and pooling layers and thus gives a feature vector for final comparison.

After introducing the basic architecture of the fundamental CNN and its functionality, in the following sections, we will explain the CNN-based 3D deep learning methods regarding different input format and targets.

7.1.3 Multiview 3D deep learning methods

In traditional computer graphics and computer vision fields, many research works have been investigated to find effective ways to convert the 3D shapes into 2D representations so that we can solve the new 3D problems by changing them into something we are familiar with in the 2D case. Thanks to this idea, the multiview image-based 3D deep learning method was proposed and achieved remarkable performance in some tough tasks. In the following part, we will introduce some 3D shape analysis methods based on CNNs for multiview 2D images.

7.1.3.1 Multiview CNN

The Multiview Convolutional Neural Networks (MVCNN) [134] is a very typical CNN using 2D images as its input to learn 3D shapes with the help of 2D image-based descriptors. The main idea of 2D multiview representations is using the 2D projections of the 3D models from different view points to express the shape as completely as possible. The Phong reflection model [110] is firstly used to make 3D shape meshes easy to render and uniformly scale them into the viewing form.

There are two kinds of image descriptors applied to each 2D view. One is based on Fisher vectors [124] with multiscale SIFT, that is, they extract the multiscale SIFT features densely from each 2D view image and then use PCA to reduce them to 80 dimensions followed by Fisher vector pooling with Gaussian mixture model. The other is CNN features, which are extracted from the penultimate layer of VGG-M network [17] (pretrained on ImageNet and fine-tuned on all 2D view images in the training dataset).

Traditional machine learning methods based on the multiview representations can be used in certain tasks to prove that this kind of inputs is very informative. The classification task adopts the one-vs-rest linear SVMs trained using the image features. In the testing process the result is obtained by selecting the highest sum among all sums of the SVM outputs for 12 views. The shape retrieval task involves the similarity criteria. In this case the distance between a shape x with n_x image descriptors and a shape y with n_y image descriptors can

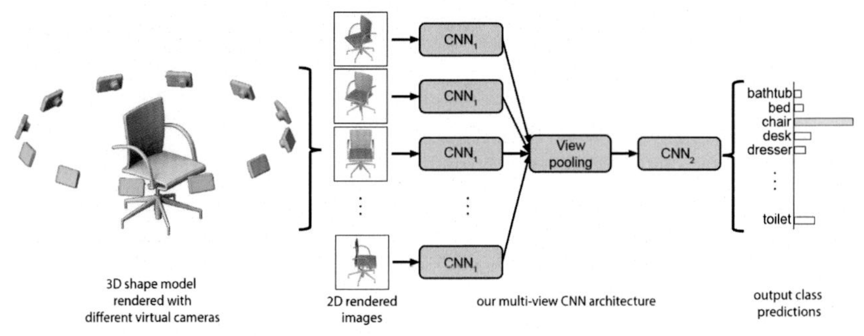

FIGURE 7.3 Multiview CNN architecture for 3D shape recognition. (Image taken from [134]. ©2015 IEEE. Included here by permission.)

be computed as

$$d(x, y) = \frac{\sum_j \min_i \|x_i - y_j\|_2}{2n_y} + \frac{\sum_i \min_j \|x_i - y_j\|_2}{2n_x}, \tag{7.1}$$

and both tasks have impressive performance.

The MVCNN architecture is shown in Fig. 7.3, which can synthesize the information from all views of a single shape into one comprehensive descriptor. At first, all n (e.g., $n = 12$ or 20) view images of a 3D shape are separately fed into n CNNs (CNN_1 in Fig. 7.3) in parallel. The n CNNs share the same weights. After this process, the output goes through the view pooling layer to perform elementwise maximum operation across the views. Then a compact descriptor aggregating all the information is sent to CNN_2 to complete the specific task.

In this way, 3D shape classification and retrieval become less burdensome. MVCNN can even do classification and retrieval based on 2D sketches. The results are better than 3D ShapeNets, which is a volumetric CNN (we will introduce later).

7.1.3.2 Projective CNN

Similar with the Multiview CNN, the Projective CNNs [56] also apply multiview 2D projection images of the query 3D object to the input layer of their networks but with different image generation strategy. In this work the 3D object mesh is firstly uniformly sampled by 1024 points on its surface. Then they select three viewpoints such that almost all sample points can be seen. In addition, for each surface sample point, the viewpoints are set to 0.5, 1.0, and 1.5 of the shape's bounding sphere radius from the sample point along its surface normal to the viewpoints. They can select a compact collection of viewpoints of different scales to well depict the 3D shape. After the viewpoints and their directions are determined, they render the query 3D shape to shaded and depth images. At each viewpoint, four viewing directions generated by four rotations along the up-vector are selected. For each direction, they get a shaded image and a

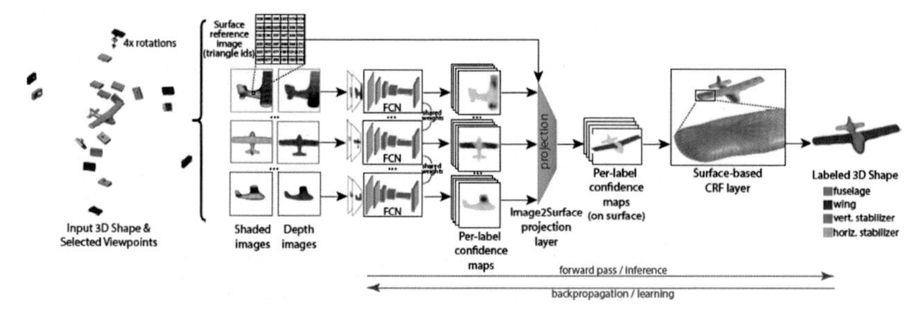

FIGURE 7.4 Pipeline of 3D shape segmentation with projective CNN. (Image taken from [56]. ©2017 IEEE. Included here by permission.)

greyscale image. Both images are then concatenated together to get a composite 2-channel image. Apart from this 2-channel image, another image, called surface reference image, is also needed, in which each pixel represents the index of a mesh polygon whose projection is closest to the pixel center.

The 2-channel images from the above operation are then sent into the image-based Fully Connected Network (FCN) modules that share the same weights as shown in Fig. 7.4. The output of each FCN is L confidence maps of size 512×512, where L is the part label number. In the following steps the output per-label confidence maps from the FCNs together with the surface reference image are fed into the Image2Surface projection layer to aggregate the multiple view information and project the processing result back onto the 3D surface. The main function of this layer is view-pooling operation, which follows Eq. (7.2) to assign a confidence $P(f, l)$ (l is a part category label) equaling to the maximum label confidence across all pixels and input images that map to polygon f according the surface reference images. In this equation, $C(m, i, j, l)$ is the confidence of label l at pixel (i, j) of image m, $I(m, i, j)$ is the polygon index at pixel (i, j), and $\tilde{C}(f, l)$ is the output confidence of label l at polygon f:

$$\tilde{C}(f, l) = \max_{m,i,j:I(m,i,j)=f} C(m, i, j, l). \tag{7.2}$$

The last step of the whole pipeline is the surface CRF (Conditional Random Field) processing, which is designed to fix the small occluded surfaces and the wrong judgment about the segmentation boundaries due to the upsampling in FCNs. In this CRF model, each polygon f has a label random variable R_f. The unary part of such a variable is defined as

$$\phi_{\text{unary}}\left(R_f = l\right) = \exp(\tilde{C}(f, l)). \tag{7.3}$$

There are also two kinds of pairwise factors, ϕ_{adj} and ϕ_{dist}; ϕ_{adj} is to encourage the polygons with the same normal to have the same part label, and ϕ_{dist} is to encourage the polygons within the small geodesic distance from each other to have the same part label. The overall CRF defined on all surface random variables $R_s = \left\{R_1, R_1, \ldots, R_{F_s}\right\}$ is

$$P(R_s) = \frac{1}{Z_s} \prod_f \phi_{\text{unary}}\left(R_f\right) \prod_{f,f'} \phi_{\text{adj}}\left(R_f, R_{f'}\right) \prod_{f,f'} \phi_{\text{dist}}\left(R_f, R_{f'}\right), \tag{7.4}$$

where Z_s is a normalizing constant.

This method combines the widely used machine learning model CRF for post-process after CNN together with a lot of data preprocessing. Although it achieves an impressive result in 3D segmentation, it shows that tackling 3D model understanding task using 2D methods is a feasible way but not a convenient way. In the following section, we will introduce the volumetric representation-based methods.

7.1.4 Volumetric 3D deep learning methods

In this section, we introduce a few 3D image-based deep learning frameworks. It is intuitive to regard the 3D model as a regular 3D image since we can easily come up with the idea to extend the deep learning processes or, more specifically, CNN operations applied on 2D pixels to 3D voxels. Quite a few researchers have demonstrated that it is indeed a feasible way. The first two networks 3D ShapeNets and VoxNet are both typical CNNs, which take 3D volumetric representations as their inputs. The 3D-GAN network combines the CNN with the powerful generative-adversarial network (GAN) to tackle the problem of generating fine synthetic 3D objects. The last CNN model is a volumetric representation input with using octree data structure to improve the computation and storage.

7.1.4.1 3D ShapeNets

The 3D ShapeNets [149] was first proposed to learn the probability distribution of complex 3D shapes across different object categories and arbitrary poses from raw CAD data. Fig. 7.5 shows the architecture of this framework.

The input of 3D ShapeNets is a $30 \times 30 \times 30$ volumetric grid bounding the 3D mesh. The voxel value is 1 if it is on or inside the mesh and 0 otherwise. To represent the probability distribution of these binary-coded 3D shapes, they combined the convolution concept in CNN with Deep Belief Networks (DBN) [40]. DBN is a powerful class of probabilistic models and is often used to model the joint probabilistic distribution over pixels and labels in 2D images. After extending this model to 3D voxels, an unavoidable problem appears: the fully connected DBN has too many parameters and becomes hard to train. As a result, the convolution is used to reduce the model parameters. However, no pooling is applied to avoid greater uncertainty for shape reconstruction.

After training, the model learns the joint distribution $p(x, y)$ of voxel data x and object category label $y \in \{1, \ldots, K\}$. Although the model is trained on

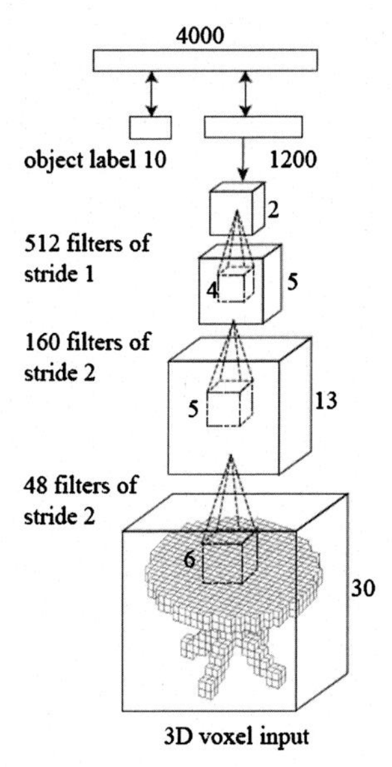

FIGURE 7.5 3D ShapeNets architecture. (Image taken from [149]. ©2015 IEEE. Included here by permission.)

complete 3D shapes, it is able to recognize the objects in single-view 2.5D depth image.

To train the 3D ShapeNet, the ModelNet, a large-scale 3D CAD model dataset, is constructed. The performance of the view-based 2.5D recognition is much better than other traditional machine learning or image processing methods.

7.1.4.2 VoxNet

Like the 3D ShapeNets that achieve impressive performance in object recognition, the VoxNet [92] makes progress on real-time object recognition, providing convenience for analyzing the 3D data from LiDAR and RGBD cameras. The VoxNet is a basic 3D CNN architecture, which is produced to create a fast and accurate object class detector for 3D point cloud data. Fig. 7.6 gives an overview of the VoxNet pipeline.

Note that before sending the data into the network, a necessary preprocess should be conducted. They get the 3D point cloud segments by extracting the intersections of the point cloud with a bounding box, followed by the procedure converting the point cloud segments into volumetric occupancy grid [98].

The occupancy grids represent the state of the environment as a 3D lattice of random variables (each corresponds to a voxel) and maintain a probabilistic estimate of their occupancy as a function of incoming sensor data and prior knowledge. This kind of representation is able to give an efficient estimate of an occupied, free, and unknown space from range measurements and is thus richer

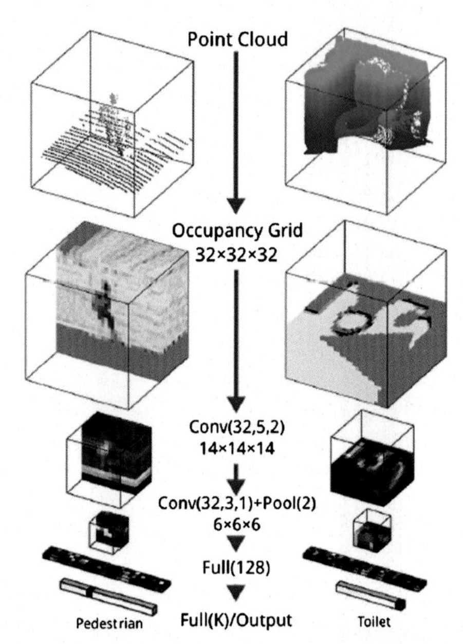

FIGURE 7.6 VoxNet pipeline. (Image taken from [92]. ©2015 IEEE. Included here by permission.)

than binary-coded 3D volumetric grid for point cloud. It can also be easily dealt with using simple data structure. The point cloud and the voxel in the grid are bijective mapping; and the origin, orientation, and resolution of the grid are researcher-defined. There are three types of occupancy models, that is, binary occupancy grid, density grid, and hit grid.

After converting the point cloud segments into $I \times J \times K$ ($I = J = K = 32$ in this case) volumetric occupancy grid, the dataset is ready to be fed into the VoxNet. The architecture of VoxNet is C(32,5,2)-C(32,3,1)-P(2)-FC(128)-FC(K), in which C(f, d, s) means a convolutional layer with f filters, each filter has the size of $d \times d \times d$ with stride s, P(m) means the pooling layer downsampling the input volume by two along each spatial dimension, FC(n) means a fully connected layer that has n hidden neurons to perform a linear combination of all outputs from the previous layer followed by a nonlinearity activation. In this VoxNet the last fully connected layer has K neurons, where K is the number of class labels.

The experiments show that the density grid representation performs the best among the three occupancy grid models and also show that the VoxNet outperforms the 3D ShapeNets in object recognition tasks greatly even using the 3D ShapeNets dataset. The experiments give us an insight that the meaningful input representation is beneficial for deep learning and that the pure CNN without grafting other neural network module is able to achieve competitive performance.

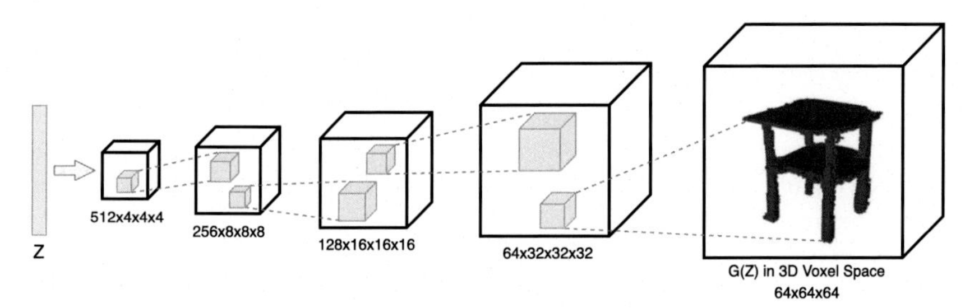

FIGURE 7.7 3D-GAN generator architecture. (Image referred from [148].)

7.1.4.3 3D generative-adversarial modeling

Unlike the 3D ShapeNets and VoxNet, the 3D Generative-Adversarial Network (3D-GAN) [148] is designed to tackle more challenging problem, 3D object generation. 3D-GAN is able to generate 3D objects from a probabilistic space by taking advantages from both volumetric CNN and generative adversarial nets (GAN).

GAN was first proposed by Goodfellow [32] and contains two main parts, a generator and a discriminator. The discriminator is trained to distinguish whether the input object is real or synthesized, whereas the generator is trained to generate synthesized object to confuse the discriminator. In 3D-GAN the generator G maps a 200-dimensional latent vector z, which is randomly sampled from a probabilistic latent space to an $n \times n \times n$ cube (here $n = 64$) containing the volumetric-represented object $G(z)$, whereas the discriminator D gives a confidence value $D(x)$ telling whether a 3D object is real or synthetic.

3D-GAN generator contains five volumetric fully convolutional layers in which the size of all filters is $4 \times 4 \times 4$ and strides are 2, with batch normalization and ReLU layers added in between and a Sigmoid layer at the end. Fig. 7.7 shows the generator architecture in 3D-GAN. The discriminator just mirrors the generator architecture with a minor replace of ReLU with Leaky ReLU [88]. No pooling or fully connected layers are implemented in 3D-GAN.

Inspired by [32], the overall adversarial loss function is

$$L_{\text{3D-GAN}} = \log D(x) + \log(1 - D(G(z))), \tag{7.5}$$

where the classification loss is implemented as a binary cross entropy loss.

The 3D-GAN can also be extended by combining an variational autoencoder (VAE) to form 3D-VAE-GAN, which acquires the ability to generate 3D object directly from the corresponding 2D image. Therefore 3D-VAE-GAN has three components: an image encoder E, a decoder with the generator G in 3D-GAN, and a discriminator D. The encoder E contains five 3D convolutional layers with filter sizes 11, 5, 5, 5, 8 and corresponding strides 4, 2, 2, 2, 1. Batch normalization and ReLU layer are also in between these convolutional layers. A 200-dimensional vector is sampled from the end and is ready to be fed into the following 3D-GAN.

Similar to VAE-GAN [74], the total loss function of 3D-VAE-GAN is

$$L = L_{\text{3D-GAN}} + \alpha_1 L_{\text{KL}} + \alpha_2 L_{\text{recon}}, \tag{7.6}$$

in which α_1 and α_2 are the coefficients of the KL divergence loss (L_{KL}) and the reconstruction loss (L_{recon}).

The experiments show the 3D-GAN is able to generate 3D objects, and 3D-VAE-GAN is able to reconstruct 3D object from the 2D images. The discriminators in GAN can do the shape classification job without supervision. All these achievements are competitive with other state-of-the-art methods.

7.1.4.4 Octree-based CNN

In the previous parts in this section, all the neural networks and the related tasks are based on volumetric representation. We can easily notice that such a kind of input format is straightforward for CNN to process, but greatly constrains the input resolution. The major challenge is that a little increase in volumetric grid dimensions would lead to a computational explosion. The O-CNN [143], an octree-based CNN, is then proposed to break the resolution bottleneck.

The breakthrough of O-CNN mainly focuses on how to represent a 3D shape in an efficient and informative way. As a result, the octree data structure is adopted to meet both requirements. In this work, they firstly construct octree representation of 3D shapes by uniformly scaling the shape into a unit 3D bounding box and then recursively subdividing the bounding box in breadth-first order. Note that in each subdivision step, only the octants occupied by the 3D shape surface are subdivided. This recursive procedure continues until the predefined subdivision depth L is reached. In this way the octants near the 3D shape surface are very small so that the fine details about the shape feature are preserved as shown in Fig. 7.8.

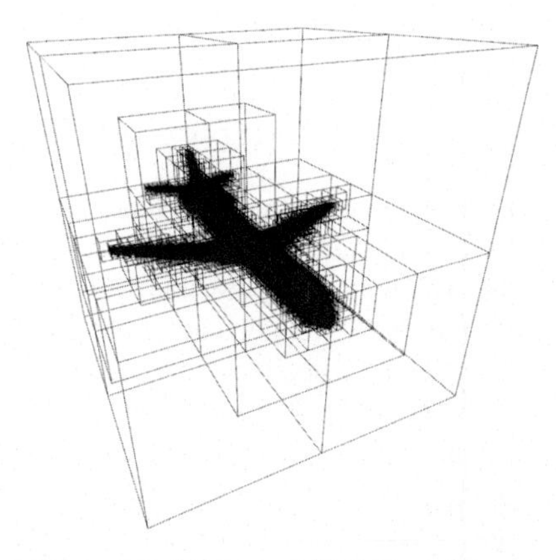

FIGURE 7.8 3D shape octree representation on an Airplane model.

According to the special octree structure of the input, the convolution operation in O-CNN can be simply realized by matrix operations [19] and computed efficiently on the GPU:

$$\Phi_c(O) = \sum_n \sum_i \sum_j \sum_k W_{ijk}^{(n)} \cdot T^{(n)}\left(O_{ijk}\right), \tag{7.7}$$

in which O_{ijk} stands for a neighboring octant of O, and T stands for the feature vector associated with O_{ijk}, $T^{(n)}$ stands for nth channel of the feature vector, and $W_{ijk}^{(n)}$ are the weights of the convolution operation. If O_{ijk} does not exist, then $T\left(O_{ijk}\right)$ is set to the zero vector.

The convolution operation with kernel size K involves the octant neighborhood searching. Instead of finding $K^3 - 1$ neighbors for a single octant, they just search $(K+1)^3 - 8$ neighboring octants for each of the 8 sibling octants. At the same time, a hash table H: $key(O)x \mapsto index(O)$ is built to facilitate the search.

Due to the special input structure, the convolution can only be performed with stride of 2^r ($r \geq 1$). The operation is applied to the first octant belonging to each sub-tree of height r, and then the feature map is downsampled by a factor of 2^r ($r \geq 1$). The pooling operation follows the similar rules.

The basic O-CNN structure consists of the units U_d in which the layer sequence is "convolution + BN (batch normalization) + ReLU + pooling". The O-CNN has the following form:

$$input \rightarrow U_d \rightarrow U_{d-1} \rightarrow \cdots \rightarrow U_2. \tag{7.8}$$

To align all features from different octree structures, they enforce all 2nd-depth octants to exist and use the zero vector padding on the empty octants at the 2nd depth.

Attaching different ends to the basic O-CNN, we can do object classification, shape retrieval, and shape part segmentation. The results outperform the traditional voxel-based CNN a lot in terms of memory and speed and are also able to compete with their performance. Adaptive O-CNN (AO-CNN) [144] is an extension of the octree-based CNNs and gains substantial efficiency in memory and computation cost compared with the existing octree-based CNNs due to the use of the adaptive octree data structure.

7.1.5 Point cloud-based 3D deep learning methods

Analyzing the geometric and semantic properties of 3D point clouds through the deep networks is still challenging due to the irregularity and sparsity of samplings of their geometric structures. The previous (multiview and volumetric) image-based methods can be conveniently developed and implemented in 3D data structure, but they easily suffer from the heavy computation and large memory expense. So it is better to define the deep learning computations based on

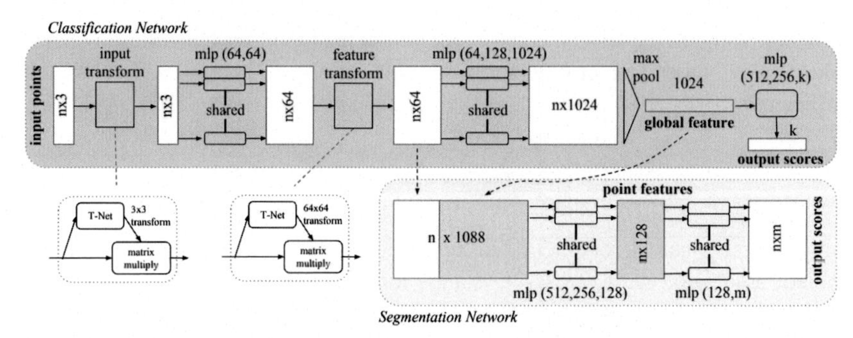

FIGURE 7.9 PointNet Architecture. The classification network takes n points as input, applies input and feature transformations, and then aggregates point features by max pooling. The output is classification scores for k classes. The segmentation network is an extension to the classification net. It concatenates global and local features and outputs per point scores. "mlp" stands for multilayer perceptron, and the numbers in bracket are layer sizes. Batchnorm is used for all layers with ReLU. Dropout layers are used for the last "mlp" in classification net. (Image taken from [111]. ©2017 IEEE. Included here by permission.)

3D shapes directly, that is, irregular/unstructured representation, such as point cloud-based methods [111,112,64,127,3,82,85,139,81,154,142,30,153]. However, defining the convolution on the irregular/unstructured representation of 3D objects is not an easy task. This section introduces a few point cloud-based deep learning frameworks.

7.1.5.1 PointNet

PointNet [111] is the first attempt of applying deep learning directly on point clouds. PointNet model is invariant to the order of points, but it considers each point independently without including local region information. Their network has three key modules: the max pooling layer as a symmetric function to aggregate information from all the points, a local and global information combination structure, and two joint alignment networks, which align both input points and point features. The full network architecture is visualized in Fig. 7.9.

Their idea is to approximate a general function defined on a point set by applying a symmetric function on transformed elements in the set:

$$f(\{x_1, \ldots, x_n\}) \approx g(h(x_1), \ldots, h(x_n)), \tag{7.9}$$

where $f : 2^{\mathbb{R}^N} \to \mathbb{R}$, $h : \mathbb{R}^N \to \mathbb{R}^K$, and $g : \underbrace{\mathbb{R}^K \times \cdots \times \mathbb{R}^K}_{n} \to \mathbb{R}$ is a symmetric function.

Empirically, their basic module is very simple: they approximate h by a multilayer perceptron network and g by a composition of a single-variable function and a max pooling function. This is found to work well by experiments. Through a collection of h, we can learn a number of fs to capture different properties of the set.

The output from the previous section forms a vector $[f_1, \ldots, f_K]$, which is a global signature of the input set. They can easily train a multilayer perceptron classifier on the shape global features for classification. After computing the global point cloud feature vector, they feed it back to per point features by concatenating the global feature with each of the point features. Then they extract new per-point features based on the combined point features, that is, the per-point feature is aware of both the local and global information. Their model can achieve shape part segmentation and scene segmentation.

The semantic labeling of a point cloud has to be invariant if the point cloud undergoes certain geometric transformations, such as a rigid transformation. We therefore expect that the learnt representation by our point set is invariant to these transformations. A natural solution is to align all input set to a canonical space before feature extraction. They predict an affine transformation matrix by a mini-network (T-net in Fig. 7.9) and directly apply this transformation to the coordinates of input points. This idea can be further extended to the alignment of feature space as well.

7.1.5.2 PointNet++

PointNet++ [112] is a hierarchical extension of PointNet model and learns local structures of point clouds at different scales.

The hierarchical structure is composed by a number of set abstraction levels as shown in Fig. 7.10. At each level, a set of points is processed and abstracted to produce a new set with fewer elements. The set abstraction level is made of three key layers: Sampling layer, Grouping layer, and PointNet layer. The Sampling layer selects a set of points from input points, which defines the centroids of local regions. Grouping layer then constructs local region sets by finding "neighboring" points around the centroids. PointNet layer uses a mini-PointNet to encode local region patterns into feature vectors. A set abstraction level takes

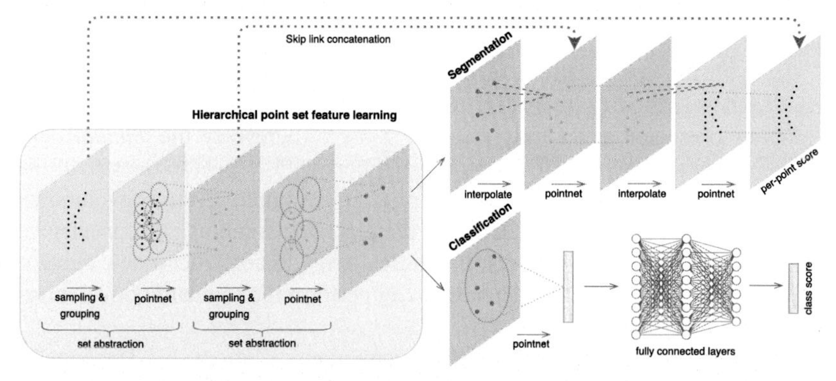

FIGURE 7.10 Illustration of our hierarchical feature learning architecture and its application for set segmentation and classification using points in 2D Euclidean space as an example. Single scale point grouping is visualized here. (Image referred from [112].)

an $N \times (d + C)$ matrix as input that is from N points with d-dim coordinates and C-dim point feature. It outputs an $N' \times (d + C')$ matrix of N' subsampled points with d-dim coordinates and new C'-dim feature vectors summarizing local context.

It is common that a point set comes with nonuniform density in different areas. Such a nonuniformity introduces a significant challenge for point set feature learning. Features learned in dense data may not generalize to sparsely sampled regions. Consequently, models trained for sparse point cloud may not recognize fine-grained local structures. To achieve this goal, they propose density adaptive PointNet layers that learn to combine features from regions of different scales when the input sampling density changes. They call this hierarchical network with density adaptive PointNet layers as PointNet++. In PointNet++, each abstraction level extracts multiple scales of local patterns and combine them intelligently according to local point densities. In terms of grouping local regions and combining features from different scales, we propose two types of density adaptive layers, that is, multiscale grouping and multiresolution grouping. Finally, they adopt a hierarchical propagation strategy with distance-based interpolation and across level skip links (Fig. 7.10).

7.1.5.3 Annularly CNNs

PointNet [111] and PointNet++ [112] consider every point in its local region independently. Some research works address the aforementioned issues by defining the convolution operator that learns the relationship between neighboring points in a local region, which helps to better capture the local geometric properties of the 3D object. [65] proposes a new end-to-end framework named as annularly convolutional neural networks (A-CNN) that leverages the neighborhood information to better capture local geometric features of 3D point clouds. The main technique components of the A-CNN model on point clouds include regular and dilated rings, constraint-based k-nearest-neighbor (k-NN) search, ordering neighbors, annular convolution, and pooling on rings. The technique details are given in [65].

Through the computations of constraint-based k-nearest-neighbor (k-NN) search and ordering neighbors as shown in Fig. 7.11A and B, the ordered neighbors are represented as an array $[\mathbf{x}_1, \mathbf{x}_2, \ldots, \mathbf{x}_K]$. To develop the *annular convolution*, they need to loop the array of neighbors with respect to the size of the kernel (e.g., 1×3, 1×5, ...) on each ring. For example, if the convolutional kernel size is 1×3, then we need to take the first two neighbors and concatenate them with the ending elements in the original array to construct a new circular array $[\mathbf{x}_1, \mathbf{x}_2, \ldots, \mathbf{x}_K, \mathbf{x}_1, \mathbf{x}_2]$. Then they can perform the standard convolutions on this array as shown in Fig. 7.11C.

There are some nice properties of the annular convolutions: (1) The annular convolution is invariant to the orientation of the local patch because the neighbors are organized and ordered in a closed loop in each ring by concatenating the beginning with the end of the sequence of neighboring points. Therefore

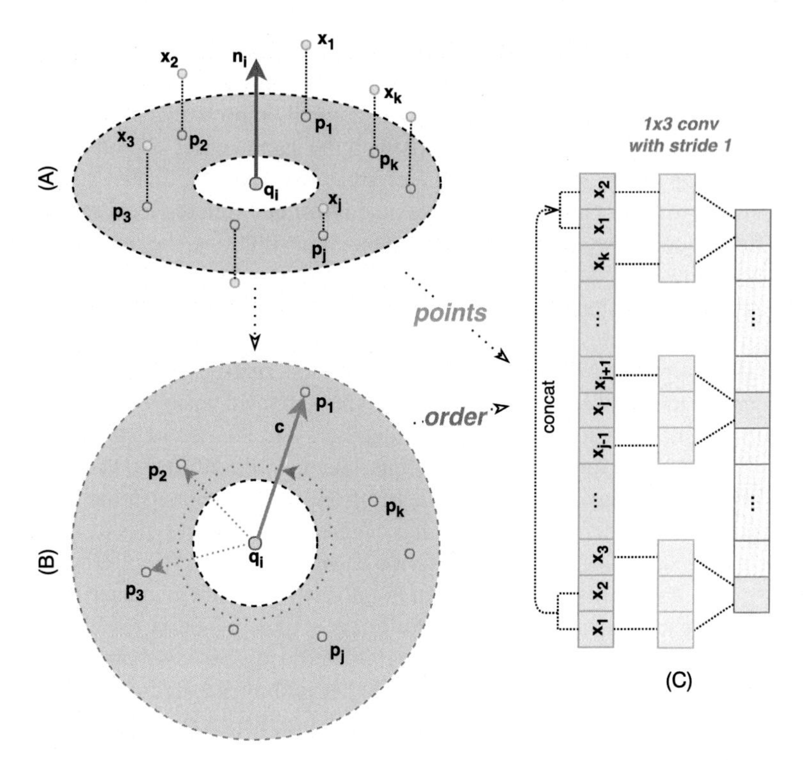

FIGURE 7.11 The illustration of the proposed annular convolution on a ring. (A) Projection: $\mathbf{q}_i$ is a query point. After applying the constraint-based k-NN search, neighboring points $\mathbf{X} = \{\mathbf{x}_j | j = 1, \ldots, K\}$ are extracted on a ring. Given the normal $\mathbf{n}_i$ at query point $\mathbf{q}_i$, the searched points are projected on the tangent plane $\mathcal{T}_i$, represented as $\mathbf{p}_j$s. (B) Counterclockwise Ordering: After projection, they randomly pick a starting point as the reference direction $\mathbf{c}$ and order points in counterclockwise. It is worth mentioning that they order original points $[\mathbf{x}_1, \mathbf{x}_2, \ldots, \mathbf{x}_j, \ldots, \mathbf{x}_K]$ based on their projections. (C) Annular Convolution: Depending on the kernel size, they copy several original points from the beginning position and concatenate them to the end of the ordered points. Finally, they apply annular convolution with the given kernel. (Image taken from [65]. ©2019 IEEE. Included here by permission.)

they can order neighbors based on any random starting position, which does not negatively affect the convolution results. Compared with some previous convolutions defined on 3D shapes [11,150,139], they all need to compute the real principal curvature direction as the reference direction to define the local patch operator, which is not robust and cumbersome. In particular, 3D shapes have large areas of flat and spherical regions, where the curvature directions are arbitrary. (2) In reality, the normal direction flipping issues are widely existing in point clouds, especially in the large-scale scene datasets. Under the annular convolution strategy, no matter the neighboring points are ordered in clockwise or counterclockwise manner, the results are the same. (3) Another advantage of annular convolution is that it can define an arbitrary kernel size, instead of just

1×1 kernels [111,112]. Therefore the annular convolution can provide the ability to learn the relationship between ordered points inside each ring as shown in Fig. 7.11C. Annular convolutions can be applied on both regular and dilated rings. By applying annular convolutions with the same kernel size on different rings they can cover and convolve larger areas by using the dilated structure, which helps to learn larger spatial contextual information in the local regions.

After sequentially applying a set of annular convolutions, the resulting convolved features encode information about its closest neighbors in each ring as well as spatial remoteness from a query point. Then they separately aggregate the convolved features across all neighbors on each ring. It applies the max-pooling strategy in this framework. The proposed ring-based scheme allows aggregating more discriminative features. The extracted max-pooled features contain the encoded information about neighbors and the relationship between them in the local region, unlike the pooling scheme in PointNet++ [112], where each neighbor is considered independently from its neighbors. In the pooling process the nonoverlapped regions (rings) aggregate different types of features in each ring, which can uniquely describe each local region (ring) around the query point. The multiscale approach in PointNet++ does not guarantee this and might aggregate the same features at different scales, which is redundant information for a network. The (regular and dilated) ring-based scheme helps to avoid extracting duplicate information but rather promotes extracting multilevel information from different regions (rings). This provides a network with more diverse features to learn from. After aggregating features at different rings, they concatenate and feed them to another abstract layer to further learn hierarchical features.

The A-CNN model follows a design where the hierarchical structure is composed of a set of abstract layers. Each abstract layer consists of several operations performed sequentially and produces a subset of input points with newly learned features. Firstly, they subsample points by using Farthest Point Sampling (FPS) algorithm [95] to extract centroids randomly distributed on the surface of each object. Secondly, the constraint-based k-NN extracts neighbors of a centroid for each local region (i.e., regular/dilated rings), and then they order neighbors in a counterclockwise manner using projection. Finally, they apply sequentially a set of annular convolutions on the ordered points and max-pool features across neighbors to produce new feature vectors, which uniquely describe each local region. Given the point clouds of 3D shapes, the proposed end-to-end network is able to classify and segment the objects.

7.2 Geometric mapping-based methods for machine learning

Manifold (mesh representation) is a quite widely used 3D shape representation. Unlike unstructured point clouds, a mesh contains the connectivity information such that it can better represent the shape of an object and preserve the geometric details. Since the mesh structure is in a nonregular and non-Euclidean format compared with the Euclidean image format, the key challenge in deep learning is how to effectively define the convolution on it. In this section, we will introduce some mesh-based CNN methods adopted in the classical 3D object understanding tasks. Note that these approaches are designed specifically for mesh learning in spatial domains.

Classically, a convolution can be considered as a template matching with filter, operating as a sliding window: for example, in an image, we extract a patch of pixels, correlate it with a template, and move the window to the next position. In the non-Euclidean setting, the lack of shift-invariance makes the patch extraction operation position-dependent. The patch operator $D_j(x)$ acting on the point $x \in \mathbf{X}$ can be defined as a reweighting of the input signal f by means of some weighting kernels $\{\omega_i(x, \cdot)\}_{i=1,\dots,J}$ spatially localized around x, that is,

$$D_j(x)f = \int_{\mathbf{X}} f(x')\omega_j(x, x')dx', \; j = 1, \dots, J. \tag{7.10}$$

The intrinsic convolution can be defined as

$$(f * g)(x) = \sum_j g_j D_j(x) f, \tag{7.11}$$

where g_j denotes the filter coefficients applied on the patch extracted at each point. Different spatial-domain intrinsic convolutional layers amount for a different definition of the patch operator D. In the following, we will introduce three related methods.

7.2.1 Directional CNNs

In [150] the authors define a directional convolution on the surface mesh with considering a robust rotation-invariant face neighboring mechanism. There are two aspects needed to be defined in the concept of the proposed *directional n-ring face neighbors*: (1) The set of n-ring face neighbors: the nth ring of a face i is the set of faces that are at distance $n-1$ from f_i in the given mesh, where the distance n is the minimum number of edges between two faces. (2) The order of n-ring face neighbors: the order of neighbors is important in convolutions since filter weights are adaptive according to their significance.

In geometry, curvatures can effectively represent the local shape variations. The local directions of minimum and maximum curvatures indicate the slowest and steepest variation of the surface normal, respectively. No matter how the model rotates, the local shape geometry is invariant. So, in this work, they choose the curvature direction as a guidance to define the order of face neighbors. For face f_i, they traverse the neighbors ring by ring based on the direction of maximum curvature counterclockwise, and the first face neighbor for each ring is the one having the minimal angle difference between two vectors, that is,

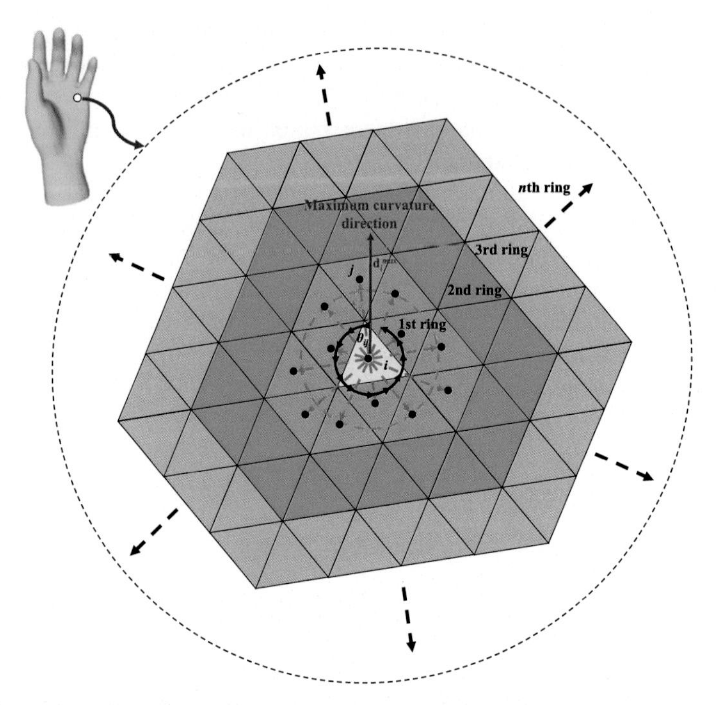

FIGURE 7.12 The illustration of the first nth rings of neighbors of face f_i under the defined order on a 3D Hand mesh model. Yellow triangle is face f_i, pink triangles are 1st ring neighbors, blue triangles are 2nd ring neighbors, brown triangles are 3rd ring neighbors, and so on. (Image taken from [150]. ©2017 IEEE. Included here by permission.)

the maximum curvature direction and the vector $\mathbf{c}_{ij}$ defined by the centroids of faces f_i and f_j. The angle between two vectors can be computed by using the geometric definition of dot product, that is, $\theta_{ij} = \cos^{-1}\left(\frac{\mathbf{d}_i^{\max}\cdot\mathbf{c}_{ij}}{\|\mathbf{d}_i^{\max}\|\cdot\|\mathbf{c}_{ij}\|}\right)$. Fig. 7.12 illustrates the first nth rings of neighbors of face f_i under the defined order on a 3D Hand mesh model.

Once directional n-ring face neighbors are defined, the definition of *directional convolution* of the feature ϕ with a kernel ω on mesh face f_i can be presented as

$$(\phi * \omega)(i) = \frac{1}{K}\sum_{j\in N_n(i)} \omega(j)\phi(j), \tag{7.12}$$

where ϕ can be a scale or vector function based on the mesh face features, such as normal, curvature, shape diameter, and so on. In this work, they only use the face normal vectors as feature, that is, $\phi(j) = \mathbf{n}_j$. The kernel ω weighs the participation of neighboring faces f_j, which will be learned during the optimization of the directional CNN, K is the normalization factor, that is,

$K = \sum_{j \in N_n(i)} \omega(j)$, $N_n(i)$ is the set of neighbors of face f_i, and n is the number of ring for face neighbors. The order of the neighbors is computed as mentioned before.

They define the filter size as $n - r \times n - r$, which means that a face and its first n rings of neighbors are convolved by the filter. If $n = 0$, then only one face is convolved by the convolution filter. Since the neighboring face number in n-ring of different faces varies, they choose the average neighbor number as filter size for n-ring and pad zeros for faces without enough neighbors (or omit redundant neighbors).

Classical pooling layers in CNN make use of the natural multiscale clustering of grid: they input all the feature maps over a cluster and output a single feature for that cluster [14]. On a surface mesh, they define a cluster as a face and its 1- to n-ring neighbors. Thus, given such a cluster, the pooling is manipulated by a downsampling strategy of a cluster of faces to 1 and denoted as $n - r \times n - r$ pooling. For max pooling, the maximum value of feature maps in the cluster is taken as the output. Similarly, the mean normal value is taken as the output for average pooling.

7.2.2 Geodesic CNNs

In [90] the authors propose to define the patch operator as a combination of Gaussian weights defined on a local intrinsic polar coordinate system. Given a point x on the shape $\mathbf{X}$, the local polar coordinate system specifies the coordinates of the surrounding points in terms of radial and angular components $(\rho(x), \theta(x))$. The radial coordinate is constructed as ρ-level sets $\{x' : d_{\mathbf{X}}(x, x') = \rho\}$ of the geodesic distance function $d_{\mathbf{X}}$ for $\rho \in [0, \rho_0]$, where ρ_0 is the radius of the geodesic disc. The angular coordinate $\theta(x)$ is constructed as a set of equispaced geodesics $\Gamma(x, \theta)$ emanating from x in direction θ in a way that they are perpendicular to the geodesic distance level sets.

Once the local geodesic coordinate system is extracted, the geodesic patch operator is defined as

$$(D(x)f)(\theta, \rho) = \int_{\mathbf{X}} f(x')\omega_\theta(x, x')\omega_\rho(x, x')dx', \tag{7.13}$$

where

$$\omega_\theta(x, x') = e^{-d^2_{\mathbf{X}}(\Gamma(x,\theta),x')/2\sigma^2_\theta}, \tag{7.14}$$

$$\omega_\rho(x, x') = e^{-(d_{\mathbf{X}}(x,x')-\rho)^2/2\sigma^2_\rho}, \tag{7.15}$$

where ω_θ and ω_ρ are the angular and radial weights, respectively (as shown in Fig. 7.13, center and right). However, note that the choice of the origin of the angular coordinate is arbitrary, and therefore it can vary from point to point. To

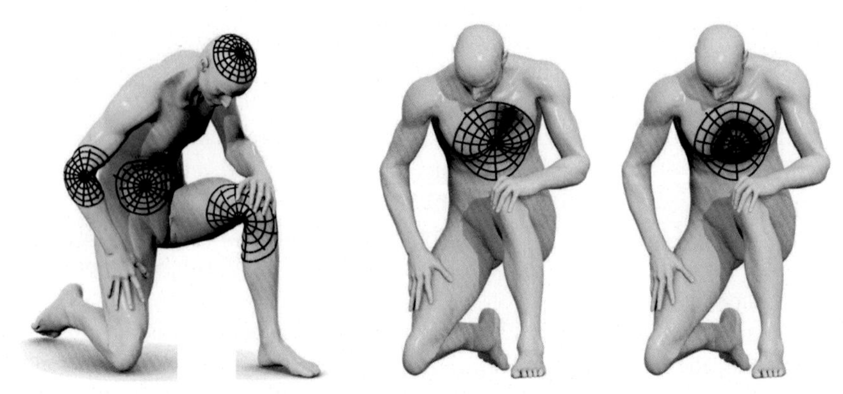

FIGURE 7.13 Construction of local geodesic polar coordinates on a manifold. Left: examples of local geodesic patches, center. Right: example of angular and radial weights, respectively (red denotes larger weights). (Image taken from [90]. ©2015 IEEE. Included here by permission.)

overcome this problem, an *angular max pooling* was used in [90], leading to the following definition of the *geodesic convolution*:

$$(f * \omega)(x) = \max_{\Delta\theta \in [0,2\pi)} \int \omega(\theta + \Delta\theta, \rho)(D(x)f)(\theta, \rho) d\theta d\rho. \tag{7.16}$$

By using the geodesic CNNs along with geometric signatures, for example, the mean curvature, conformal factor, heat kernel, and so on, we can well extract and express the actual local and global geometric features for deep learning-based shape applications, such as learning facial expressions with 3D mesh CNNs [54].

7.2.3 Anisotropic CNNs

The heat propagation on a shape $\mathbf{X}$ is governed by the heat diffusion equation. In particular, given the delta function centered on x as an initial heat distribution, the heat distribution on $\mathbf{X}$ after some time t is represented by the heat kernel $h_t(x, \cdot)$. An isotropic heat kernel diffuses heat equally in all directions (Fig. 7.14, left), and the diffusion strength is controlled by the diffusion time t.

An *anisotropic* heat diffusion is described by the anisotropic diffusion equation

$$f_t(x, t) = -\text{div}(\mathbf{A}(x)\nabla f(x, t)), \tag{7.17}$$

where the thermal conductivity matrix $\mathbf{A}(x)$ specifies the heat conductivity properties at each point on the shape $\mathbf{X}$ and contains both positional and directional information. This more general diffusion model was considered in [1] for shape analysis tasks. In [11] the authors defined the thermal conductivity

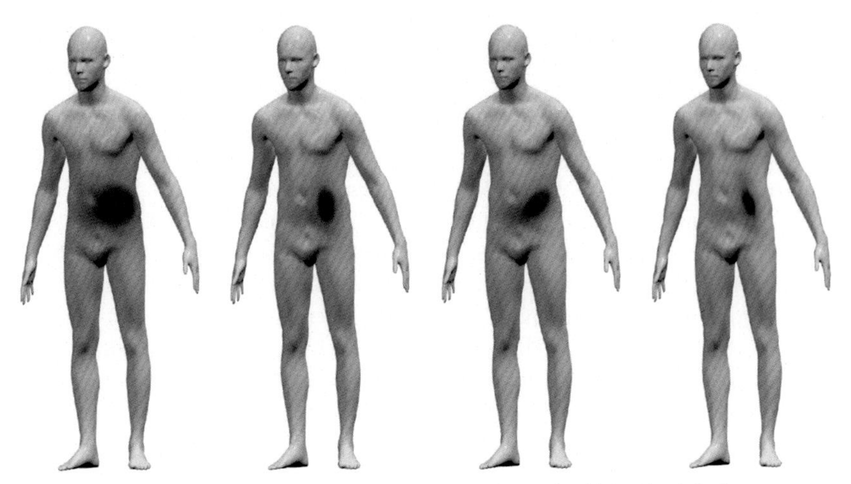

FIGURE 7.14 Visualization of different heat kernels (red represent high values). Leftmost: example of an isotropic heat kernel. Remaining: examples of anisotropic heat kernels for different rotation angles θ and anisotropy coefficient α. (Image taken from [91]. ©2016 ACM, Inc. Included here by permission.)

matrix as

$$\mathbf{A}_{\alpha\theta}(x) = \mathbf{R}_{\theta}(x)\text{diag}(\alpha, 1)\mathbf{R}_{\theta}(x)^T, \tag{7.18}$$

where the matrix $\mathbf{R}_{\theta}(x)$ performs rotation of θ with respect to some reference (e.g., the maximum curvature) direction, and $\alpha > 0$ is a parameter controlling the magnitude of anisotropy. Particularly, $\alpha = 1$ corresponds to the classical isotropic case.

Analogously to the heat diffusion in the Fourier domain, the anisotropic heat kernel is given by

$$h_{\alpha\theta t}(x, x') = \sum_{k\geq 0} e^{-t\lambda_{\alpha\theta,k}} \phi_{\alpha\theta,k}(x)\phi_{\alpha\theta,k}(x'), \tag{7.19}$$

where $\phi_{\alpha\theta,k}(x)$ and $\lambda_{\alpha\theta,k}$ are the eigenfunctions and eigenvalues of the anisotropic Laplacian $\Delta_{\alpha\theta} = -\text{div}(\mathbf{A}_{\alpha\theta}(x)\nabla)$. The anisotropic heat kernel $h_{\alpha\theta t}$ depends on two additional parameters, the coefficient α and the rotation angle θ. Fig. 7.14 shows some examples of anisotropic heat kernels computed at different rotations θ and anisotropies α. In [11], such kernels are used as the weighting functions for the construction of patch operator in Eq. (7.10):

$$(D(x)f)(\theta, t) = \int_{\mathbf{X}} h_{\alpha\theta t}(x, x')f(x')dx', \tag{7.20}$$

mapping the values of f around point x to a local polar-like system of coordinates (θ, t).

7.2.4 Deformation CNNs

Recently, there is an emerging trend to derive and analyze the 3D/4D shape from the single-view 2D images by deep neural network, such as using the deformation-based CNNs on 3D shapes driving from 2D images. In general, these methods can be divided into two major categories, surface deformation and space deformation. The *surface deformation* is directly defined on the shape surface to predict a nice surface space, whose computational effort and numerical robustness are highly related to the complexity and quality of the surface [12]. For example, in [55], it presents a novel approach to analyze the facial expressions from 2D images through learning of a 3D morphable face model and a quantitative information visualization scheme for exploring this type of visual data. From an input image a 3D face with expression can be reconstructed by using the deformation CNNs based on some prior knowledge, from which basis parameters and a displacement map are extracted as features for facial emotion analysis and visualization. Based upon the features, two support vector regressions are trained to determine the fuzzy valence-arousal (VA) values to quantify the emotions. The emotion analysis and visualization system based on a sequence of the reconstructed 3D facial shapes can detect expressions robustly from various head poses, face sizes, and lighting conditions and is fully automatic to compute the VA values from images or a sequence of video with various facial expressions.

Compared with the surface-based deformation methods, *space deformation* approaches apply a trivariate deformation function to transform all the points of the original surface. The key idea is to deform the ambient space (i.e., 3D volume space) enclosing the shapes and thus to implicitly deform the embedded surface shape (i.e., 2-manifold) [12]. One major advantage of the space deformation is that it does not depend on any particular surface representation, so that it can be used to deform all kinds of explicit surface representations, such as vertices of meshes or samples of point clouds. The classical free-form deformation (FFD) represents the space deformation by a trivariate tensor-product spline function [126]. Recently, there is a deep neural network-based method, that is, DeepOrganNet [145], to generate and visualize fully high-fidelity 3D/4D organ geometric models from single-view medical images with complicated background in real time. This organ reconstruction network simultaneously learns the optimal selection and the best smooth deformation from multiple templates via a trivariate tensor-product deformation technique, that is, FFD, to match the query 2D image. It proposes an end-to-end deep learning method with a lightweight and effective neural network to reconstruct multiple high-fidelity 3D organ meshes with a variety of geometric shapes from a single-view medical image with complicated background and noises. Finally, it can reconstruct and analyze both the 3D and 4D (i.e., a sequence of 3D) lung shapes and their motions in real time. The major contributions of this work are to accurately reconstruct the 3D/4D organ shapes from 2D single-view projection, significantly improve the procedure time to allow on-the-fly visualization, and dramatically reduce the imaging dose for human subjects.

7.3 Spectral geometric features for machine learning

In the following sections, we introduce 3D geometric deep learning approaches designed specifically in spectral domains. We first need compute the spectral descriptors on each point of a 3D model. These spectral descriptors are then fed into the proposed spectral CNNs as an input. In this section, we describe multiple spectral descriptors that are used as inputs for neural networks.

7.3.1 Global points signature

In [121], given a point on a 2-manifold, the global point signature (GPS) at the point x is defined as

$$\mathrm{GPS}(x) = \left[\frac{\phi_1(x)}{\sqrt{\lambda_1}}, \frac{\phi_2(x)}{\sqrt{\lambda_2}}, \ldots, \frac{\phi_n(x)}{\sqrt{\lambda_n}} \right]^T, \tag{7.21}$$

where λ_k and ϕ_k are the kth eigenvalue and eigenfunction of the Laplace–Beltrami operator defined on the manifold, respectively. A suggested number of eigenvalues in [121] is $n = 25$.

7.3.2 Heat kernel signature

Sun et al. [136] proposed a construction of intrinsic dense descriptors by considering the diagonal of the heat kernel, known as the *autodiffusivity* function:

$$h_t(x, x) = \sum_{k \geq 0} e^{-t\lambda_k} \phi_k^2(x). \tag{7.22}$$

The physical interpretation of autodiffusivity is the amount of heat remaining at point x after time t. Geometrically, autodiffusivity is related to the Gaussian curvature. Sun et al. [136] defined the heat kernel signature (HKS) of dimension Q at point x by sampling the autodiffusivity function at some fixed times $t_1, \ldots, t_Q$:

$$\mathbf{f}(x) = (h_{t_1}(x, x), \ldots, h_{t_Q}(x, x))^T. \tag{7.23}$$

The HKS is a very popular approach in numerous applications due to some appealing properties: (1) It is intrinsic and hence invariant to isometric deformations of the manifold by construction. (2) It is dense. (3) The spectral expression in Eq. (7.22) of the heat kernel allows efficient computation of the HKS by using the first few eigenvectors and eigenvalues of the Laplace–Beltrami operator. At the same time, a drawback of HKS stemming from the use of low-pass filters is poor spatial localization.

7.3.3 Wave kernel signature

Aubry et al. [4] considered a physical model of a quantum particle on the manifold by the Schrödinger equation:

$$(i\Delta \mathbf{x} + \frac{\partial}{\partial t})\psi(x,t) = 0, \tag{7.24}$$

where $\psi(x,t)$ is the complex wave function capturing the particle behavior. Assuming that the particle oscillates at frequency λ drawn from a probability distribution $\pi(\lambda)$, the solution of Eq. (7.24) can be expressed in the Fourier domain as

$$\psi(x,t) = \sum_{k\geq 1} e^{i\lambda_k t}\pi(\lambda_k)\phi_k(x). \tag{7.25}$$

The probability of finding the particle at point x is given by

$$p(x) = \lim_{T\to\infty}\int_0^T |\psi(x,t)|^2 dt = \sum_{k\geq 1}\pi^2(\lambda_k)\phi_k^2(x) \tag{7.26}$$

and depends on the initial frequency distribution $\pi(\lambda)$. Aubry et al. [4] considered a log-normal frequency distribution $\pi_\nu(\lambda) = e^{(\frac{\log\nu - \log\lambda}{2\sigma^2})}$ with mean frequency ν and standard deviation σ. They defined the Q-dimensional wave kernel signature (WKS)

$$\mathbf{f}(x) = (p_{\nu_1}(x), \ldots, p_{\nu_Q}(x))^T, \tag{7.27}$$

where $p_\nu(x)$ is the probability in Eq. (7.26) corresponding to the initial log-normal frequency distribution with mean frequency ν, and $\nu_1, \ldots, \nu_Q$ are some logarithmically sampled frequencies.

Note that WKS is based on log-normal transfer functions that act as band-pass filters and thus exhibits better spatial localization.

7.3.4 Optimal spectral descriptors

Litman and Bronstein [84] considered generic descriptors of the form

$$\mathbf{f}(x) = \sum_{k\geq 1}\tau(\lambda_k)\phi_k^2(x) \approx \sum_{k=1}^{K}\tau(\lambda_k)\phi_k^2(x), \tag{7.28}$$

where $\tau(\lambda) = (\tau_1(\lambda), \ldots, \tau_Q(\lambda))^T$ is a bank of transfer functions acting on LBO eigenvalues, and used the parametric transfer functions

$$\tau_q(\lambda) = \sum_{m=1}^{M}\alpha_{qm}\beta_m(\lambda) \tag{7.29}$$

in the B-spline basis $\beta_1(\lambda), \ldots, \beta_M(\lambda)$, where $\alpha_{qm}(q = 1, \ldots, Q,\ m = 1, \ldots, M)$ are the parameterization coefficients. Inserting Eq. (7.29) into Eq. (7.28), we can express the qth component of the spectral descriptor as

$$f_q(x) = \sum_{k \geq 1} \tau_q(\lambda_k)\phi_k^2(x) = \sum_{m=1}^{M} \alpha_{qm} g_m(x) \tag{7.30}$$

and

$$g_m(x) = \sum_{k \geq 1} \beta_m(\lambda_k)\phi_k^2(x), \tag{7.31}$$

where $\mathbf{g}(x) = (g_1(x), \ldots, g_M(x))^T$ is a vector-valued function referred to as geometry vector dependent only on the intrinsic geometry of the shape. Thus Eq. (7.28) is parameterized by the $Q \times M$ matrix $\mathbf{A} = (\alpha_{lm})$ and can be written in matrix form as $\mathbf{f}(x) = \mathbf{A}\mathbf{g}(x)$. The main idea of [84] is to learn the optimal parameters $\mathbf{A}$ by minimizing a task-specific loss that reduces to a Mahalanobis-type metric learning.

7.4 Learning of spectral geometry with convolutional neural networks

The convolution $f * g$ of two functions in the spectral domain can be expressed as the elementwise product of their Fourier transforms:

$$\widehat{f * g}(\omega) = \hat{f}(\omega) \cdot \hat{g}(\omega). \tag{7.32}$$

Eq. (7.32) can be used to define convolution on non-Euclidean domains in the spectral domain as

$$f * g(x) = \sum_{k \geq 0} \langle f, \phi_k \rangle_{L^2(\mathbf{x})} \langle g, \phi_k \rangle_{L^2(\mathbf{x})} \phi_k(x). \tag{7.33}$$

Such an operation can be interpreted as a nonlinear filtering of the signal f. The key difference from the classical convolution is the lack of shift-invariance, which makes the filter kernel to change depending on its position.

7.4.1 Spectral CNNs

In [14] the spectral definition of convolution in Eq. (7.33) was used to generalized CNNs to graphs by representing filters in the spectral domain. The fundamental drawback of this formulation is its limitation to a single given domain, since the spectral representation of the filters is basis dependent. It implies that if we learn a filter with respect to a basis on one domain and then apply it on another domain with another basis, then the result can be very different as shown in Fig. 7.15.

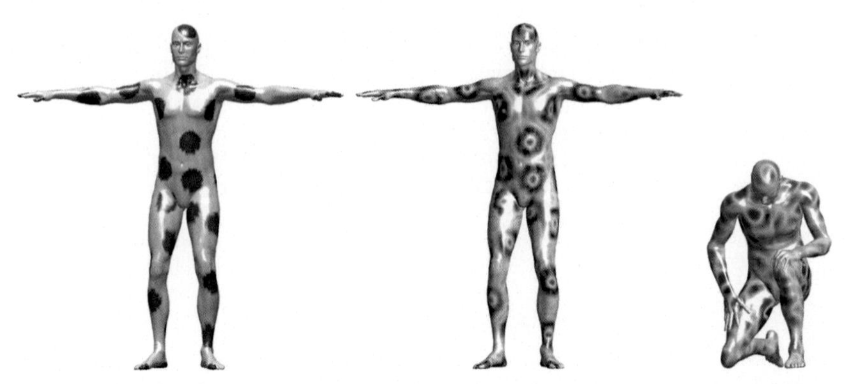

FIGURE 7.15 An illustration of the poor generalization of spectral filtering across non-Euclidean domains. Left: a function defined on a manifold. Middle: A result of the application of a filter in the frequency domain on the same manifold. Right: the same filter applied to the same function but on a different (nearly isometric) domain produces a completely different result. (Image taken from [13]. ©2017 IEEE. Included here by permission.)

7.4.2 Localized spectral CNNs

One of the main drawbacks of the Fourier transform is that it is global, that is, the basis functions have a global support. The localized or *windowed Fourier transform (WFT)* is used in signal processing for local space-frequency analysis of a signal. The main idea of the WFT is to analyze the frequency content of a signal that is localized by means of multiplication by a window. Given a function $f \in L^2(\mathbb{R})$ and some "mother window" g localized at zero, the classical WFT is defined as

$$(Sf)(x, \omega) = \int_{-\infty}^{+\infty} f(x')g(x - x')e^{-ix'\omega}dx'. \tag{7.34}$$

It can be also defined as an inner product with a translated and modulated window:

$$(Sf)(x, \omega) = \langle f, M_\omega T_x g \rangle_{L^2(\mathbb{R})}, \tag{7.35}$$

where T_x and M_ω denote the translation and modulation operators, respectively. The translated and modulated window $M_\omega T_x g$ is referred to as the WFT atom. In the Euclidean setting, the translation operator is defined simply as $(T_{x'} f)(x) = f(x - x')$, whereas the modulation operator is a multiplication by a Laplacian eigenfunction $(M_\omega f)(x) = e^{i\omega x} f(x)$, which amounts to a translation in the frequency domain $(\widehat{M_{\omega'} f}) = \hat{f}(\omega - \omega')$.

The translation $x - x'$ to a point x' is not well defined in the non-Euclidean setting. In [133,10], these operations were defined in the frequency domain. Translation to x' is replaced by convolution in Eq. (7.33) with a delta-function

FIGURE 7.16 Examples of different WFT atoms $g_{x,k}$ using different windows (top and bottom rows; window Fourier coefficients are shown on the left), shown in different localizations (second and third columns) and modulations (fourth and fifth columns). (Image taken from [10]. ©2015 John Wiley & Sons, Inc. Included here by permission.)

centered at x', yielding:

$$\begin{aligned}(T_{x'} f)(x) = (f * \delta'_x)(x) &= \sum_{k\geq 1} \langle f, \phi_k \rangle_{L^2(\mathbf{X})} \langle \delta'_x, \phi_k \rangle_{L^2(\mathbf{X})} \phi_k(x) \\ &= \sum_{k\geq 1} \langle f, \phi_k \rangle_{L^2(\mathbf{X})} \phi_k(x') \phi_k(x). \end{aligned} \tag{7.36}$$

Note that such a translation is not shift-invariant in general, that is, the window changes when moved around the manifold (Fig. 7.16). The modulation operator is defined as $(M_k f)(x) = \phi_k(x) f(x)$, where ϕ_k is the kth eigenfunction of the Laplace–Beltrami operator. Combining the two operators together, the WFT atom (examples as shown in Fig. 7.16) becomes

$$g_{x',k}(x) = (M_k T_{x'} g)(x) = \phi_k(x) \sum_{i\geq 1} \hat{g}_i \phi_i(x) \phi_i(x'). \tag{7.37}$$

Note that the "mother window" is defined here in the frequency domain by the coefficients $\hat{g}_i$. Finally, the WFT of a signal $f \in L^2(\mathbf{X})$ can be defined as

$$(Sf)(x', k) = \langle f, g_{x',k} \rangle_{L^2(\mathbf{X})} = \sum_{i\geq 1} \hat{g}_i \phi_i(x') \langle f, \phi_i \phi_k \rangle_{L^2(\mathbf{X})}. \tag{7.38}$$

The WFT $(Sf)(x', k)$ performs a filtering of the signal f at the point x' at the frequency k. By collecting its behavior over different frequencies the content of the signal f in a local support around x' is extracted, reproducing in this way the window extraction on images. The *localized spectral convolution layer* can be defined as

$$\mathbf{g}_l(x') = \sum_{l'=1}^{p} \sum_{k=1}^{K} \omega_{l,k,l'} |(S\mathbf{f}_{l'})(x', k)|, \tag{7.39}$$

where $\mathbf{f}_{l'}, l' = 1, \ldots, p$, is the input signal, $W = \omega_{l,k,l'}$ is a $p \times K \times q$ tensor representing the learnable weights, and $\mathbf{g}_l, l = 1, \ldots, q$, is the output signal. An additional degree of freedom is the possibility to learn the window itself as well.

A drawback of this approach is its memory and computation requirements, as each window $\hat{g}_i$ in Eq. (7.38) needs to be explicitly produced.

7.4.3 Synchronized spectral CNNs

The basic architecture of synchronized spectral CNNs (SyncSpecCNN) is similar to the fully convolutional segmentation network in [86]. They repeat the operation of convolving the vertex function by kernels and applying nonlinear transformation. However, they have several key differences. First, SyncSpecCNN achieves multichannel convolution by multichannel modulation in the spectral domain. Second, SyncSpecCNN parameterizes kernels in the spectral domain following a dilated fashion, so that kernel sizes could be effectively enlarged to capture large context information without increasing the number of parameters. This is essentially a spectral counterpart of spatial pooling in [86]. Last, they design a Spectral Transformer Network to synchronize the spectral domain of different shapes, allowing better parameter sharing. More details are given in [154].

7.5 Hands-on experiments and applications on shape classification with spectral geometry CNNs

The following web links are providing some source codes and hands-on experiments from the original authors of the aforementioned papers to help us getting familiar with the discussed methods and algorithms. We are grateful to their courtesies.

Multiview Convolutional Neural Networks (MVCNN): the source code is available at https://github.com/suhangpro/mvcnn.

3D Shape Segmentation with Projective Convolutional Networks: the source code is available at https://github.com/kalov/ShapePFCN.

3D ShapeNets: the source code is available at https://github.com/zhirongw/3DShapeNets.

VoxNet: the source code is available at https://github.com/dimatura/voxnet.

3D Generative-Adversarial Network (3D-GAN): the source code is available at https://github.com/zck119/3dgan-release.

O-CNN: Octree-based Convolutional Neural Networks: the source code is available at https://github.com/Microsoft/O-CNN.

PointNet: the source code is available at https://github.com/charlesq34/pointnet.

PointNet++: the source code is available at https://github.com/charlesq34/pointnet2.

A-CNN: the source code is available at https://github.com/artemkomarichev/a-cnn.

Directional CNNs: the source code is available at https://github.com/HaotianMXu/3D-Shape-Segmentation-with-Deep-Neural-Networks.

Localized Spectral CNNs: the source code is available at https://github.com/jonat hanmasci/ShapeNet.

Synchronized Spectral CNN for 3D Shape Segmentation: the source code is available at https://github.com/ericyi/SyncSpecCNN.

7.6 Summary

The recent emergence of geometric deep learning methods in various communities and application domains triggers a new research direction. In this chapter, we have introduced the current progress achieved in 3D geometric deep learning area. Among all the 3D deep learning approaches, image-based methods are the most well-developed. By adopting such strategies, we can deal with not only the 3D models of volumetric representations, but also the 3D models with multiview images and point cloud-based representations. Then we discussed several mesh-based CNNs methods in spatial domains, such as directional CNNs, geodesic CNNs, anisotropic diffusion CNNs, and deformation CNNs. After that, we introduced some spectral geometric descriptors. Finally, we discussed the global, local, and synchronized spectral geometric CNN methods. We provided some corresponding source codes of the discussed methods.

Chapter 8

Conclusion

Shape analysis is a fundamental research topic in computer graphics and computer vision including matching, retrieval, mapping, and so on. Inspired by the recent research, we focus on the shapes represented with differential geometry as the differential operators contain the intrinsic geometry information of the original shape. On one hand, the second-order Laplace–Beltrami operator introduces a spectral domain where the Euclidean transformations and isometric deformations are filtered out. In such a spectral domain, only the intrinsic shape properties are left. This book has focused on the shape analysis based on such geometry behind the differential operators.

The main contributions include:

- We have introduced a novel 3D shape representation with a set of salient feature points in the Laplace–Beltrami spectrum. The Laplace–Beltrami operator is defined on the geometry of a Riemannian manifold. The Laplace eigenvalue problem introduces spectra of shape geometry similar to the Fourier transfer on time domain. They also have the similar properties. The shape spectra depend on shape geometry and are invariant to translation, rotation, scaling, and isometric deformations. The shape is then projected to the spectral bases and represented with a linear combination of them. The salient features are extracted from the "frequency" domain as local geometry energy maxima, which shared the invariability of the original spectrum. The maxima provide not only where the features are on the manifold but also the "frequency" where the features lie in, so the scales of the salient features are predicted. With spectra, shapes are represented with a set of invariant salient features with scales. IQP is employed to retrieve the correspondences among variant shapes in very detailed levels. The experiment results show the applications in shape matching, retrieval, and searching. Partial matching is also supported in our framework.
- We have presented a novel method to understand the poses in the geometry spectral domain. Poses are defined as a set of near-isometric shapes casted by the same model. Their Laplace–Beltrami spectra stay stable under minor nonisometry deformations. All the poses from the same object can be reembedded to a high-dimensional spectral domain. By transferring spatial poses into the spectral domain their geometry are aligned naturally there despite Euclidean transformations, triangulations, and near-isometric deformations. In this case the pose differences are represented with local geometry properties, for example, mean curvatures. Investigating in spectral domain, the pose

Spectral Geometry of Shapes. https://doi.org/10.1016/B978-0-12-813842-7.00016-4

motions will vary the local properties on the fixed geometry. Large variations indicate joints, and small ones rigid parts. The eigenfunction also carry rich geometric meaning, which leads to an automatic skeleton extraction. Combining the part understanding and the skeleton, the semantic deformable model is obtained. The shape spectra help shape understandings and segmentations. It contributes to motion analysis in computer vision and pattern recognition tasks as well.

- We have proved that the shape spectrum is a piecewise analytic to a scale function on the conformal factor on the Riemannian metric of the manifold. The derivatives of the eigenvalues are expressed with those of the scale function at each time. The property applies to both continuous domain and discrete triangle meshes. Furthermore, a spectrum alignment algorithm is developed on the triangle meshes. In the discrete domain, integration is represented with matrix product. The derivatives of the shape spectrum and the scale vector can be turned into a matrix form, which introduces a linear system. We have applied the smoothness and local bound constraints to solve the linear system by reaching the minimum energy of a quadratic programming problem. Given two closed triangle meshes representing manifolds, the eigenvalues can be aligned from one to the other. After the eigenvalues are registered, the eigenfunction distributions are aligned as well. This means that the shape spectrum can be controlled by the user analytically with a scale vector and nonisometric deformation analysis is available within shape spectra.
- We have presented a novel surface registration technique using the spectrum of the shapes, which can facilitate accurate localization and visualization of nonisometric deformations of the surfaces. To register two surfaces, we map both eigenvalues and eigenvectors of the Laplace–Beltrami operator of the shapes through optimizing an energy function. The function is defined by the integration of a smoothness term to align the eigenvalues and a distance term between the eigenvectors at feature points to align the eigenvectors. The feature points are generated using the static points of certain eigenvectors of the surfaces. By using both the eigenvalues and the eigenvectors on these feature points, the computational efficiency is improved considerably without losing the accuracy in comparison to the approaches that use the eigenvectors for all vertices. In our technique the variation of the shape is expressed using a scale function defined at each vertex. Consequently, the total energy function to align the two given surfaces can be defined using the linear interpolation of the scale function derivatives. Through the optimization of the energy function, the scale function can be solved, and the alignment is achieved. After the alignment, the eigenvectors can be employed to calculate the point-to-point correspondence of the surfaces. Therefore the proposed method can accurately define the displacement of the vertices. We have evaluated the method by conducting experiments on synthetic and real data using hippocampus, heart, and hand models.

- We have described the main mathematical ideas behind geometric deep learning. With these materials, we have provided a clear picture of the key concepts and techniques and illustrated the related applications. The deep learning-based geometry is a data-driven methodology and can be conceptually considered as data geometry. Among all the 3D deep learning approaches, image-based methods are the most well-developed. Using such strategies, we can deal with not only the 3D models of volumetric representations, but also the 3D models with multiview images, as well as point cloud-based representations. Then several mesh-based CNNs methods in spatial domains, such as directional CNNs, geodesic CNNs, anisotropic diffusion CNNs, and deformation CNNs, have been discussed. Some spectral geometric descriptors have been also introduced, along with the global, local, and synchronized spectral geometric CNN methods. We have also provided development details for several applications in shape analysis and synthesis, especially in spectral geometry.

Bibliography

[1] Mathieu Andreux, Emanuele Rodola, Mathieu Aubry, Daniel Cremers, Anisotropic Laplace–Beltrami operators for shape analysis, in: Proceedings of the European Conference on Computer Vision, 2014, pp. 299–312.

[2] Mihael Ankerst, Gabi Kastenmüller, Hans-Peter Kriegel, Thomas Seidl, 3D shape histograms for similarity search and classification in spatial databases, in: Proceedings of the International Symposium on Advances in Spatial Databases, 1999.

[3] Matan Atzmon, Haggai Maron, Yaron Lipman, Point convolutional neural networks by extension operators, ACM Transactions on Graphics 37 (4) (2018) 71.

[4] Mathieu Aubry, Ulrich Schlickewei, Daniel Cremers, The wave kernel signature: a quantum mechanical approach to shape analysis, in: Proceedings of the IEEE International Conference on Computer Vision Workshops, 2011, pp. 1626–1633.

[5] Rine Bakkestrom, Nicolaj L. Christensen, Emil Wolsk, Ann Banke, Jordi S. Dahl, Mads J. Andersen, Finn Gustafsson, Christian Hassager, Jacob E. Moller, Layer-specific deformation analysis in severe aortic valve stenosis, primary mitral valve regurgitation, and healthy individuals validated against invasive hemodynamic measurements of heart function, Echocardiography (2018) 1–9.

[6] Alberto Bemporad, Domenico Mignone, Manfred Morari, An efficient branch and bound algorithm for state estimation and control of hybrid systems, in: Proceedings of the European Control Conference, 1999, pp. 557–562.

[7] Mirela Ben-Chen, Craig Gotsman, Guy Bunin, Conformal flattening by curvature prescription and metric scaling, Computer Graphics Forum 27 (2) (2008) 449–458.

[8] Paul J. Besl, Neil D. McKay, A method for registration of 3D shapes, IEEE Transactions on Pattern Analysis and Machine Intelligence 14 (2) (1992).

[9] Dmitriy Bespalov, William C. Regli, Ali Shokoufandeh, Reeb graph based shape retrieval for CAD, in: Proceedings of the ASME Design Engineering Technical Conferences, 2003, pp. 2–6.

[10] Davide Boscaini, Jonathan Masci, Simone Melzi, Michael M. Bronstein, Umberto Castellani, Pierre Vandergheynst, Learning class-specific descriptors for deformable shapes using localized spectral convolutional networks, Computer Graphics Forum 34 (2015) 13–23.

[11] Davide Boscaini, Jonathan Masci, Emanuele Rodolà, Michael Bronstein, Learning shape correspondence with anisotropic convolutional neural networks, in: Proceedings of the Advances in Neural Information Processing Systems, 2016, pp. 3189–3197.

[12] Mario Botsch, Leif Kobbelt, Mark Pauly, Pierre Alliez, Bruno Lévy, Polygon Mesh Processing, AK Peters/CRC Press, 2010.

[13] Michael M. Bronstein, Joan Bruna, Yann LeCun, Arthur Szlam, Pierre Vandergheynst, Geometric deep learning: going beyond Euclidean data, IEEE Signal Processing Magazine 34 (4) (2017) 18–42.

[14] Joan Bruna, Wojciech Zaremba, Arthur Szlam, Yann LeCun, Spectral networks and locally connected networks on graphs, in: Proceedings of the International Conference on Learning Representations, 2014.

[15] Hamish Carr, Jack Snoeyink, Ulrike Axen, Computing contour trees in all dimensions, in: Proceedings of the ACM-SIAM Symposium on Discrete Algorithms, 2000, pp. 918–926.
[16] Hamish Carr, Jack Snoeyink, Michiel van de Panne, Simplifying flexible isosurfaces using local geometric measures, in: Proceedings of the Conference on Visualization, 2004, pp. 497–504.
[17] Ken Chatfield, Karen Simonyan, Andrea Vedaldi, Andrew Zisserman, Return of the devil in the details: delving deep into convolutional nets, preprint, arXiv:1405.3531, 2014.
[18] Chavel Issac, Eigenvalues in Riemannian Geometry, Pure and Applied Mathematics, Academic Press, 1984.
[19] Kumar Chellapilla, Sidd Puri, Patrice Simard, High performance convolutional neural networks for document processing, in: Proceedings of the International Workshop on Frontiers in Handwriting Recognition, 2006.
[20] Ding-Yun Chen, Ming Ouhyoung, A 3D object retrieval system based on multi-resolution Reeb graph, in: Proceedings of the Computer Graphics Workshop, 2002, pp. 16–20.
[21] Hung-Kuo Chu, Tong-Yee Lee, Multiresolution mean shift clustering algorithm for shape interpolation, IEEE Transactions on Visualization and Computer Graphics 15 (5) (2009) 853–866.
[22] Kree Cole-McLaughlin, Herbert Edelsbrunner, John Harer, Vijay Natarajan, Valerio Pascucci, Loops in Reeb graphs of 2-manifolds, in: Proceedings of the Symposium on Computational Geometry, 2003, pp. 344–350.
[23] Nicu D. Cornea, Deborah Silver, Patrick Min, Curve-skeleton properties, applications, and algorithms, IEEE Transactions on Visualization and Computer Graphics 13 (3) (2007) 530–548.
[24] Luca Cosmo, Mikhail Panine, Arianna Rampini, Maks Ovsjanikov, Michael M. Bronstein, Emanuele Rodolà, Isospectralization, or how to hear shape, style, and correspondence, CoRR, arXiv:1811.11465, 2018.
[25] Tamal K. Dey, Pawas Ranjan, Yusu Wang, Convergence, stability, and discrete approximation of Laplace spectra, in: Proceedings of the ACM-SIAM Symposium on Discrete Algorithms, 2010, pp. 650–663.
[26] Richard O. Duda, Peter E. Hart, Pattern Classification and Scene Analysis, Wiley, 1973.
[27] Mohamed El-Mehalawi, R. Allen Miller, A database system of mechanical components based on geometric and topological similarity. Part I: representation, Computer Aided Design 35 (2003) 83–94.
[28] Mohamed El-Mehalawi, R. Allen Miller, A database system of mechanical components based on geometric and topological similarity. Part II: indexing, retrieval, matching, and similarity assessment, Computer Aided Design 35 (2003) 95–105.
[29] Michael Elad, Ayellet Tal, Sigal Ar, Content based retrieval of VRML objects: an iterative and interactive approach, in: Proceedings of the Eurographics Workshop on Multimedia, 2002, pp. 107–118.
[30] Francis Engelmann, Theodora Kontogianni, Alexander Hermans, Bastian Leibe, Exploring spatial context for 3D semantic segmentation of point clouds, preprint, arXiv:1801.00470, 2018.
[31] Ran Gal, Daniel Cohen-Or, Salient geometric features for partial shape matching and similarity, ACM Transactions on Graphics 25 (1) (2006) 130–150.
[32] Ian Goodfellow, Jean Pouget-Abadie, Mehdi Mirza, Bing Xu, David Warde-Farley, Sherjil Ozair, Aaron Courville, Yoshua Bengio, Generative adversarial nets, in: Proceedings of the Advances in Neural Information Processing Systems, 2014, pp. 2672–2680.
[33] Xianfeng Gu, Steven J. Gortler, Hugues Hoppe, Geometry images, ACM Transactions on Graphics 21 (3) (2002) 355–361.
[34] Xianfeng Gu, Shing-Tung Yau, Global conformal surface parameterization, in: Proceedings of the 2003 Eurographics/ACM SIGGRAPH Symposium on Geometry Processing, 2003, pp. 127–137.
[35] André P. Guéziec, Xavier Pennec, Nicholas Ayache, Medical image registration using geometric hashing, IEEE Computational Science & Engineering 4 (4) (1997) 29–41.

[36] Hajar Hamidian, Jiaxi Hu, Zichun Zhong, Jing Hua, Quantifying shape deformations by variation of geometric spectrum, in: Proceedings of the International Conference on Medical Image Computing and Computer-Assisted Intervention, 2016, pp. 150–157.
[37] Hajar Hamidian, Zichun Zhong, Farshad Fotouhi, Jing Hua, Surface registration with eigenvalues and eigenvectors, IEEE Transactions on Visualization and Computer Graphics (2019).
[38] Ying He, Xian Xiao, Hock-Soon Seah, Harmonic 1-form based skeleton extraction from examples, Graphical Models 71 (2) (2009) 49–62.
[39] Masaki Hilaga, Yoshihisa Shinagawa, Taku Kohmura, Tosiyasu L. Kunii, Topology matching for fully automatic similarity estimation of 3D shapes, in: Proceedings of the Annual Conference on Computer Graphics and Interactive Techniques, 2001, pp. 203–212.
[40] Geoffrey E. Hinton, Simon Osindero, Yee-Whye Teh, A fast learning algorithm for deep belief nets, Neural Computation 18 (7) (2006) 1527–1554.
[41] Berthold K.P. Horn, Extended Gaussian images, Proceedings of the IEEE 72 (2) (1984) 1671–1686.
[42] Jiaxi Hu, Shape Analysis Using Spectral Geometry, Wayne State University Dissertations, 2015.
[43] Jiaxi Hu, Hajar Hamidian, Zichun Zhong, Jing Hua, Visualizing shape deformations with variation of geometric spectrum, IEEE Transactions on Visualization and Computer Graphics 23 (1) (2017) 721–730.
[44] Jiaxi Hu, Jing Hua, Salient spectral geometric features for shape matching and retrieval, The Visual Computer 25 (5–7) (2009) 667–675.
[45] Jiaxi Hu, Jing Hua, Pose analysis using spectral geometry, The Visual Computer 29 (9) (2013) 949–958.
[46] Jing Hua, Zhaoqiang Lai, Ming Dong, Xianfeng Gu, Hong Qin, Geodesic distance-weighted shape vector image diffusion, IEEE Transactions on Visualization and Computer Graphics 14 (6) (2008) 1643–1650.
[47] Wyke Huizinga, Dirk H.J. Poot, Meike W. Vernooij, Gennady V. Roshchupkin, Esther E. Bron, Mohammad Arfan Ikram, Daniel Rueckert, Wiro J. Niessen, Stefan Klein, Alzheimer's Disease Neuroimaging Initiative, A spatio-temporal reference model of the aging brain, NeuroImage 169 (2018) 11–22.
[48] Bulat Ibragimov, Tomaz Vrtovec, Landmark-Based Statistical Shape Representations, Elsevier, 2017.
[49] Cheuk Yiu Ip, Daniel Lapadat, Leonard Sieger, William C. Regli, Using shape distributions to compare solid models, in: Proceedings of the ACM Symposium on Solid Modeling and Applications, 2002, pp. 273–280.
[50] Natraj Iyer, Yagnanarayanan Kalyanaraman, Kuiyang Lou, Subramaniam Jayanti, Karthik Ramani, A reconfigurable 3D engineering shape search system. Part I: shape representation, in: Proceedings of the ASME Computers and Information in Engineering Conference, 2003.
[51] Varun Jain, Hao Zhang, Robust 3D shape correspondence in the spectral domain, in: Proceedings of the IEEE International Conference on Shape Modeling and Applications, 2006, pp. 19–19.
[52] Doug L. James, Christopher D. Twigg, Skinning mesh animations, ACM Transactions on Graphics 24 (3) (2005) 399–407.
[53] Tao Jiang, Kun Qian, Shuang Liu, Jing Wang, Xiaosong Yang, Jianjun Zhang, Consistent as-similar-as-possible non-isometric surface registration, The Visual Computer 33 (6) (2017) 891–901.
[54] Hai Jin, Yuanfeng Lian, Jing Hua, Learning facial expressions with 3D mesh convolutional neural network, ACM Transactions on Intelligent Systems and Technology 10 (1) (2018) 7.
[55] Hai Jin, Xun Wang, Yuanfeng Lian, Jing Hua, Emotion information visualization through learning of 3D morphable face model, The Visual Computer 35 (4) (2019) 535–548.
[56] Evangelos Kalogerakis, Melinos Averkiou, Subhransu Maji, Siddhartha Chaudhuri, 3D shape segmentation with projective convolutional networks, in: Proceedings of the IEEE Conference on Computer Vision and Pattern Recognition, 2017, pp. 3779–3788.

[57] Zachi Karni, Craig Gotsman, Spectral compression of mesh geometry, in: Proceedings of the Annual Conference on Computer Graphics and Interactive Techniques, 2000, pp. 279–286.

[58] Zachi Karni, Craig Gotsman, Compression of soft-body animation sequences, Computers & Graphics 28 (1) (2004) 25–34.

[59] Michael Kazhdan, Thomas Funkhouser, Harmonic 3D shape matching, in: Proceedings of the ACM SIGGRAPH Conference Abstracts and Applications, 2002, pp. 191–191.

[60] Michael Kazhdan, Thomas Funkhouser, Szymon Rusinkiewicz, Rotation invariant spherical harmonic representation of 3D shape descriptors, in: Proceedings of the Eurographics/ACM SIGGRAPH Symposium on Geometry Processing, 2003, pp. 156–164.

[61] David G. Kendall, The diffusion of shape, Advances in Applied Probability 9 (3) (1977) 428–430.

[62] Liliya Kharevych, Boris Springborn, Peter Schröder, Discrete conformal mappings via circle patterns, ACM Transactions on Graphics 25 (2) (2006) 412–438.

[63] Martin Kilian, Niloy J. Mitra, Helmut Pottmann, Geometric modeling in shape space, ACM Transactions on Graphics 26 (3) (2007).

[64] Roman Klokov, Victor Lempitsky, Escape from cells: deep Kd-networks for the recognition of 3D point cloud models, in: Proceedings of the IEEE International Conference on Computer Vision, 2017, pp. 863–872.

[65] Artem Komarichev, Zichun Zhong, Jing Hua, A-CNN: annularly convolutional neural networks on point clouds, in: Proceedings of the IEEE Conference on Computer Vision and Pattern Recognition, 2019.

[66] Ender Konukoglu, Ben Glocker, Antonio Criminisi, Kilian M. Pohl, WESD–weighted spectral distance for measuring shape dissimilarity, IEEE Transactions on Pattern Analysis and Machine Intelligence 35 (9) (2013) 2284–2297.

[67] Artiom Kovnatsky, Michael M. Bronstein, Alexander M. Bronstein, Klaus Glashoff, Ron Kimmel, Coupled quasi-harmonic bases, Computer Graphics Forum 32 (2) (2013) 439–448.

[68] Artiom Kovnatsky, Michael M. Bronstein, Xavier Bresson, Pierre Vandergheynst, Functional correspondence by matrix completion, in: Proceedings of the IEEE Conference on Computer Vision and Pattern Recognition, 2015, pp. 905–914.

[69] Alex Krizhevsky, Ilya Sutskever, Geoffrey E. Hinton, ImageNet classification with deep convolutional neural networks, in: Proceedings of the Advances in Neural Information Processing Systems, 2012, pp. 1097–1105.

[70] Rongjie Lai, Yonggang Shi, Ivo Dinov, Tony F. Chan, Arthur W. Toga, Laplace–Beltrami nodal counts: a new signature for 3D shape analysis, in: Proceedings of the IEEE International Symposium on Biomedical Imaging: From Nano to Macro, 2009, pp. 694–697.

[71] Zhaoqiang Lai, Jiaxi Hu, Chang Liu, Vahid Taimouri, Darshan Pai, Jiong Zhu, Jianrong Xu, Jing Hua, Intra-patient supine-prone colon registration in CT colonography using shape spectrum, in: Tianzi Jiang, Nassir Navab, Josien P.W. Pluim, Max A. Viergever (Eds.), Proceedings of the International Conference on Medical Image Computing and Computer-Assisted Intervention, 2010, pp. 332–339.

[72] Zhaoqiang Lai, Jing Hua, 3D surface matching and registration through shape images, in: Proceedings of the International Conference on Medical Image Computing and Computer-Assisted Intervention, 2008, pp. 44–51.

[73] Yehezkel Lamdan, Haim J. Wolfson, Geometric hashing: a general and efficient model based recognition scheme, in: Proceedings of the IEEE International Conference on Computer Vision, 1988, pp. 238–249.

[74] Anders Boesen Lindbo Larsen, Søren Kaae Sønderby, Hugo Larochelle, Ole Winther, Autoencoding beyond pixels using a learned similarity metric, preprint, arXiv:1512.09300, 2015.

[75] Yann LeCun, Léon Bottou, Yoshua Bengio, Patrick Haffner, Gradient-based learning applied to document recognition, Proceedings of the IEEE 86 (11) (1998) 2278–2324.

[76] Chang Ha Lee, Amitabh Varshney, David W. Jacobs, Mesh saliency, ACM Transactions on Graphics 24 (3) (2005) 659–666.

[77] Nathaniel Leibowitz, Zipora Y. Fligelman, Ruth Nussinov, Haim J. Wolfson, Multiple structural alignment and core detection by geometric hashing, in: Proceedings of the International Conference on Intelligent Systems for Molecular Biology, 1999, pp. 169–177.

[78] Marc Levoy, Kari Pulli, Brian Curless, Szymon Rusinkiewicz, David Koller, Lucas Pereira, Matt Ginzton, Sean Anderson, James Davis, Jeremy Ginsberg, Jonathan Shade, Duane Fulk, The digital Michelangelo project: 3D scanning of large statues, in: Proceedings of the Annual Conference on Computer Graphics and Interactive Techniques, 2000, pp. 131–144.

[79] Bruno Lévy, Laplace–Beltrami eigenfunctions towards an algorithm that "understands" geometry, in: Proceedings of the IEEE International Conference on Shape Modeling and Applications, 2006, pp. 13–13.

[80] Bruno Lévy, Sylvain Petitjean, Nicolas Ray, Jérome Maillot, Least squares conformal maps for automatic texture atlas generation, ACM Transactions on Graphics 21 (3) (2002) 362–371.

[81] Jiaxin Li, Ben M. Chen, Gim Hee Lee, SO-Net: self-organizing network for point cloud analysis, in: Proceedings of the IEEE Conference on Computer Vision and Pattern Recognition, 2018, pp. 9397–9406.

[82] Yangyan Li, Rui Bu, Mingchao Sun, Wei Wu, Xinhan Di, Baoquan Chen, PointCNN: convolution on X-transformed points, in: Proceedings of the Advances in Neural Information Processing Systems, 2018, pp. 828–838.

[83] Or Litany, Emanuele Rodolà, Alexander M. Bronstein, Michael M. Bronstein, Fully spectral partial shape matching, Computer Graphics Forum 36 (2) (2017) 247–258.

[84] Roee Litman, Alexander M. Bronstein, Learning spectral descriptors for deformable shape correspondence, IEEE Transactions on Pattern Analysis and Machine Intelligence 36 (1) (2014) 171–180.

[85] Xinhai Liu, Zhizhong Han, Yu-Shen Liu, Matthias Zwicker, Point2Sequence: learning the shape representation of 3D point clouds with an attention-based sequence to sequence network, in: Proceedings of the Association for the Advancement of Artificial Intelligence, vol. 33, 2019, pp. 8778–8785.

[86] Jonathan Long, Evan Shelhamer, Trevor Darrell, Fully convolutional networks for semantic segmentation, in: Proceedings of the IEEE Conference on Computer Vision and Pattern Recognition, 2015, pp. 3431–3440.

[87] Kuiyang Lou, Subramaniam Jayanti, Natraj Iyer, Yagnanarayanan Kalyanaraman, Sunil Prabhakar, Karthik Ramani, A reconfigurable 3D engineering shape search system. Part II: database indexing, retrieval and clustering, in: Proceedings of the ASME Computers and Information in Engineering Conference, 2003, pp. 169–178.

[88] Andrew L. Maas, Awni Y. Hannun, Andrew Y. Ng, Rectifier nonlinearities improve neural network acoustic models, in: Proceedings of the International Conference on Machine Learning, vol. 30 (1), 2013, pp. 3–3.

[89] David Marr, Vision: a Computational Investigation into the Human Representation and Processing of Visual Information, W.H. Freeman, 1982.

[90] Jonathan Masci, Davide Boscaini, Michael Bronstein, Pierre Vandergheynst, Geodesic convolutional neural networks on Riemannian manifolds, in: Proceedings of the IEEE International Conference on Computer Vision Workshops, 2015, pp. 37–45.

[91] Jonathan Masci, Emanuele Rodolà, Davide Boscaini, Michael M. Bronstein, Hao Li, Geometric Deep Learning, SIGGRAPH ASIA Course Notes, 2016.

[92] Daniel Maturana, Sebastian Scherer, VoxNet: a 3D convolutional neural network for real-time object recognition, in: Proceedings of the IEEE/RSJ International Conference on Intelligent Robots and Systems, 2015, pp. 922–928.

[93] Russell Merris, Laplacian matrices of graphs: a survey, Linear Algebra and Its Applications 197 (1994) 143–176.

[94] Mark Meyer, Mathieu Desbrun, Peter Schröder, Alan Barr, Discrete differential geometry operators for triangulated 2-manifolds, in: Visualization and Mathematics III, Springer, 2003, pp. 35–57.

[95] Carsten Moenning, Neil A. Dodgson, Fast marching farthest point sampling, Technical report, University of Cambridge, Computer Laboratory, 2003.
[96] Bojan Mohar, The Laplacian spectrum of graphs, Graph Theory, Combinatorics, and Applications, Springer (1991) 871–898.
[97] Bojan Mohar, Laplace eigenvalues of graphs—a survey, Discrete Mathematics 109 (1–3) (1992) 171–183.
[98] Hans Moravec, Alberto Elfes, High resolution maps from wide angle sonar, in: Proceedings of the IEEE International Conference on Robotics and Automation, vol. 2, 1985, pp. 116–121.
[99] Kentaro Mori, Atsushi Yukawa, Atsushi Kono, Yutaka Hata, Heart failure diagnosis for tagged magnetic resonance images, in: Proceedings of the International Conference on Machine Learning and Cybernetics, vol. 1, 2017, pp. 67–70.
[100] Alexander Naitsat, Shichao Cheng, Xiaofeng Qu, Xin Fan, Emil Saucan, Yehoshua Y. Zeevi, Geometric approach to detecting volumetric changes in medical images, Journal of Computational and Applied Mathematics 329 (2018) 37–50.
[101] Marcin Novotni, Reinhard Klein, 3D Zernike descriptors for content based shape retrieval, in: Proceedings of the ACM Symposium on Solid Modeling and Applications, 2003, pp. 216–225.
[102] Ryutarou Ohbuchi, Masatoshi Nakazawa, Tsuyoshi Takei, Retrieving 3D shapes based on their appearance, in: Proceedings of the ACM SIGMM International Workshop on Multimedia Information Retrieval, 2003, pp. 39–45.
[103] Ryutarou Ohbuchi, Tsuyoshi Takei, Shape-similarity comparison of 3D models using alpha shapes, in: Proceedings of the Pacific Conference on Computer Graphics and Applications, 2003, pp. 293–302.
[104] Ozan Oktay, Wenjia Bai, Ricardo Guerrero, Martin Rajchl, Antonio de Marvao, Declan P. O'Regan, Stuart A. Cook, Mattias P. Heinrich, Ben Glocker, Daniel Rueckert, Stratified decision forests for accurate anatomical landmark localization in cardiac images, IEEE Transactions on Medical Imaging 36 (1) (2017) 332–342.
[105] Robert Osada, Thomas Funkhouser, Bernard Chazelle, David Dobkin, Shape distributions, ACM Transactions on Graphics 21 (4) (2002) 807–832.
[106] Maks Ovsjanikov, Mirela Ben-Chen, Frederic Chazal, Leonidas J. Guibas, Analysis and visualization of maps between shapes, Computer Graphics Forum 32 (6) (2013) 135–145.
[107] Maks Ovsjanikov, Mirela Ben-Chen, Justin Solomon, Adrian Butscher, Leonidas Guibas, Functional maps: a flexible representation of maps between shapes, ACM Transactions on Graphics 31 (4) (2012) 30.
[108] Valerio Pascucci, Giorgio Scorzelli, Peer-Timo Bremer, Ajith Mascarenhas, Robust on-line computation of Reeb graphs: simplicity and speed, ACM Transactions on Graphics 26 (3) (2007).
[109] Giuseppe Patanè, Michela Spagnuolo, Bianca Falcidieno, A minimal contouring approach to the computation of the Reeb graph, IEEE Transactions on Visualization and Computer Graphics 15 (4) (2009) 583–595.
[110] Bui Tuong Phong, Illumination for computer generated pictures, Communications of the ACM 18 (6) (1975) 311–317.
[111] Charles R. Qi, Hao Su, Kaichun Mo, Leonidas J. Guibas, PointNet: deep learning on point sets for 3D classification and segmentation, in: Proceedings of the IEEE Conference on Computer Vision and Pattern Recognition, 2017, pp. 652–660.
[112] Charles Ruizhongtai Qi, Li Yi, Hao Su, Leonidas J. Guibas, PointNet++: deep hierarchical feature learning on point sets in a metric space, in: Proceedings of the Advances in Neural Information Processing Systems, 2017, pp. 5099–5108.
[113] P. Radau, Y. Lu, K. Connelly, G. Paul, A. Dick, G. Wright, Evaluation framework for algorithms segmenting short axis cardiac MRI, The MIDAS Journal - Cardiac MR Left Ventricle Segmentation Challenge (2009).
[114] William C. Regli, Vincent A. Cicirello, Managing digital libraries for computer-aided design, Computer Aided Design 32 (2) (2000) 119–132.

[115] Martin Reuter, Marc Niethammer, Franz-Erich Wolter, Sylvain Bouix, Martha Shenton, Global medical shape analysis using the volumetric Laplace spectrum, in: Proceedings of the International Conference on Cyberworlds, NASA-GEM Workshop, 2007, pp. 417–426.

[116] Martin Reuter, Franz-Erich Wolter, Niklas Peinecke, Laplace–Beltrami spectra as "shape-DNA" of surfaces and solids, Computer Aided Design 38 (4) (2006) 342–366.

[117] Martin Reuter, Laplace Spectra for Shape Recognition, Books on Demand, 2006.

[118] Martin Reuter, Hierarchical shape segmentation and registration via topological features of Laplace–Beltrami eigenfunctions, International Journal of Computer Vision 89 (2–3) (2010) 287–308.

[119] Martin Reuter, Franz-Erich Wolter, Martha Shenton, Marc Niethammer, Laplace–Beltrami eigenvalues and topological features of eigenfunctions for statistical shape analysis, Computer Aided Design 41 (10) (2009) 739–755.

[120] Emanuele Rodolà, Luca Cosmo, Michael M. Bronstein, Andrea Torsello, Daniel Cremers, Partial functional correspondence, Computer Graphics Forum 36 (1) (2017) 222–236.

[121] Raif M. Rustamov, Laplace–Beltrami eigenfunctions for deformation invariant shape representation, in: Proceedings of the Eurographics Symposium on Geometry Processing, 2007, pp. 225–233.

[122] Raif M. Rustamov, Maks Ovsjanikov, Omri Azencot, Mirela Ben-Chen, Frédéric Chazal, Leonidas Guibas, Map-based exploration of intrinsic shape differences and variability, ACM Transactions on Graphics 32 (4) (2013) 72.

[123] Firooz A. Sadjadi, Ernest L. Hall, Three-dimensional moment invariants, IEEE Transactions on Pattern Analysis and Machine Intelligence 2 (2) (1980) 127–136.

[124] Jorge Sánchez, Florent Perronnin, Thomas Mensink, Jakob Verbeek, Image classification with the Fisher vector: theory and practice, International Journal of Computer Vision 105 (3) (2013) 222–245.

[125] Dietmar Saupe, Dejan V. Vranic, 3D model retrieval with spherical harmonics and moments, in: Proceedings of the DAGM-Symposium on Pattern Recognition, 2001, pp. 392–397.

[126] Thomas W. Sederberg, Scott R. Parry, Free-form deformation of solid geometric models, ACM SIGGRAPH Computer Graphics 20 (4) (1986) 151–160.

[127] Yiru Shen, Chen Feng, Yaoqing Yang, Dong Tian, Mining point cloud local structures by kernel correlation and graph pooling, in: Proceedings of the IEEE Conference on Computer Vision and Pattern Recognition, 2018, pp. 4548–4557.

[128] Yonggang Shi, Rongjie Lai, Raja Gill, Daniel Pelletier, David Mohr, Nancy Sicotte, Arthur W. Toga, Conformal metric optimization on surface (CMOS) for deformation and mapping in Laplace–Beltrami embedding space, in: Proceedings of the International Conference on Medical Image Computing and Computer-Assisted Intervention, 2011, pp. 327–334.

[129] Yonggang Shi, Rongjie Lai, Arthur W. Toga, Cortical surface reconstruction via unified Reeb analysis of geometric and topological outliers in magnetic resonance images, IEEE Transactions on Medical Imaging 32 (3) (2013) 511–530.

[130] Yonggang Shi, Rongjie Lai, Danny J.J. Wang, Daniel Pelletier, David Mohr, Nancy Sicotte, Arthur W. Toga, Metric optimization for surface analysis in the Laplace–Beltrami embedding space, IEEE Transactions on Medical Imaging 33 (7) (2014) 1447–1463.

[131] Yoshihisa Shinagawa, Tosiyasu L. Kunii, Constructing a Reeb graph automatically from cross sections, IEEE Computer Graphics and Applications 11 (6) (1991) 44–51.

[132] Heung-Yeung Shum, Martial Hebert, Katsushi Ikeuchi, On 3D shape similarity, in: Proceedings of the Conference on Computer Vision and Pattern Recognition, 1996, pp. 526–531.

[133] David I. Shuman, Benjamin Ricaud, Pierre Vandergheynst, Vertex-frequency analysis on graphs, Applied and Computational Harmonic Analysis 40 (2) (2016) 260–291.

[134] Hang Su, Subhransu Maji, Evangelos Kalogerakis, Erik Learned-Miller, Multi-view convolutional neural networks for 3D shape recognition, in: Proceedings of the IEEE International Conference on Computer Vision, 2015, pp. 945–953.

[135] Robert W. Sumner, Jovan Popović, Deformation transfer for triangle meshes, ACM Transactions on Graphics 23 (3) (2004) 399–405.

[136] Jian Sun, Maks Ovsjanikov, Leonidas Guibas, A concise and provably informative multi-scale signature based on heat diffusion, Computer Graphics Forum 28 (2009) 1383–1392.
[137] Hari Sundar, Deborah Silver, Nikhil Gagvani, Sven Dickinson, Skeleton based shape matching and retrieval, in: Proceedings of the Shape Modeling International, 2003, pp. 130–139.
[138] Vahid Taimouri, Jing Hua, Visualization of shape motions in shape space, IEEE Transactions on Visualization and Computer Graphics 19 (12) (2013) 2644–2652.
[139] Maxim Tatarchenko, Jaesik Park, Vladlen Koltun, Qian-Yi Zhou, Tangent convolutions for dense prediction in 3D, in: Proceedings of the IEEE Conference on Computer Vision and Pattern Recognition, 2018, pp. 3887–3896.
[140] Tony Tung, Francis Schmittt, Shape retrieval of noisy watertight models using aMRG, in: Proceedings of the International Conference on Shape Modeling and Applications, 2008, pp. 229–230.
[141] Dejan V. Vranic, Dietmar Saupe, Jörg Richter, Tools for 3D object retrieval: Karhunen–Loeve transform and spherical harmonics, in: Proceedings of the IEEE Workshop on Multimedia Signal Processing, 2001, pp. 293–298.
[142] Chu Wang, Babak Samari, Kaleem Siddiqi, Local spectral graph convolution for point set feature learning, in: Proceedings of the European Conference on Computer Vision, 2018, pp. 52–66.
[143] Peng-Shuai Wang, Yang Liu, Yu-Xiao Guo, Chun-Yu Sun, Xin Tong, O-CNN: octree-based convolutional neural networks for 3D shape analysis, ACM Transactions on Graphics 36 (4) (2017) 72.
[144] Peng-Shuai Wang, Chun-Yu Sun, Yang Liu, Xin Tong, Adaptive O-CNN: a patch-based deep representation of 3D shapes, ACM Transactions on Graphics 37 (6) (2018) 217.
[145] Yifan Wang, Zichun Zhong, Jing Hua, DeepOrganNet: on-the-fly reconstruction and visualization of 3D/4D lung models from single-view projections by deep deformation network, IEEE Transactions on Visualization and Computer Graphics (2019).
[146] Ofir Weber, Olga Sorkine, Yaron Lipman, Craig Gotsman, Context-aware skeletal shape deformation, Computer Graphics Forum 26 (3) (2007) 265–274.
[147] Haim J. Wolfson, Isidore Rigoutsos, Geometric hashing: an overview, IEEE Computational Science & Engineering 4 (4) (1997) 10–21.
[148] Jiajun Wu, Chengkai Zhang, Tianfan Xue, Bill Freeman, Josh Tenenbaum, Learning a probabilistic latent space of object shapes via 3D generative-adversarial modeling, in: Proceedings of the Advances in Neural Information Processing Systems, 2016, pp. 82–90.
[149] Zhirong Wu, Shuran Song, Aditya Khosla, Fisher Yu, Linguang Zhang, Xiaoou Tang, Jianxiong Xiao, 3D ShapeNets: a deep representation for volumetric shapes, in: Proceedings of the IEEE Conference on Computer Vision and Pattern Recognition, 2015, pp. 1912–1920.
[150] Haotian Xu, Ming Dong, Zichun Zhong, Directionally convolutional networks for 3D shape segmentation, in: Proceedings of the IEEE International Conference on Computer Vision, 2017, pp. 2698–2707.
[151] Dong-Ming Yan, Bruno Lévy, Yang Liu, Feng Sun, Wenping Wang, Isotropic remeshing with fast and exact computation of restricted Voronoi diagram, Computer Graphics Forum 28 (5) (2009) 1445–1454.
[152] Han-Bing Yan, Shimin Hu, Ralph R. Martin, Yong-Liang Yang, Shape deformation using a skeleton to drive simplex transformations, IEEE Transactions on Visualization and Computer Graphics 14 (3) (2008) 693–706.
[153] Xiaoqing Ye, Jiamao Li, Hexiao Huang, Liang Du, Xiaolin Zhang, 3D recurrent neural networks with context fusion for point cloud semantic segmentation, in: Proceedings of the European Conference on Computer Vision, 2018, pp. 403–417.
[154] Li Yi, Hao Su, Xingwen Guo, Leonidas Guibas, SyncSpecCNN: synchronized spectral CNN for 3D shape segmentation, in: Proceedings of the Computer Vision and Pattern Recognition, 2017, pp. 2282–2290.
[155] Yang Yu, Shaoting Zhang, Junzhou Huang, Dimitris Metaxas, Leon Axel, Sparse deformable models with application to cardiac motion analysis, in: Proceedings of the International Conference on Information Processing in Medical Imaging, vol. 7917, 2013, pp. 208–219.

[156] Wei Zeng, Lok Ming Lui, Lin Shi, Defeng Wang, Winnie C.W. Chu, Jack C.Y. Cheng, Jing Hua, Shing-Tung Yau, Xianfeng Gu, Shape analysis of vestibular systems in adolescent idiopathic scoliosis using geodesic spectra, in: Proceedings of the International Conference on Medical Image Computing and Computer-Assisted Intervention, 2010, pp. 538–546.

[157] Zichun Zhong, Xiaohu Guo, Wenping Wang, Bruno Lévy, Feng Sun, Yang Liu, Weihua Mao, Particle-based anisotropic surface meshing, ACM Transactions on Graphics 32 (4) (2013) 99.

[158] Kun Zhou, John Synder, Baining Guo, Heung-Yeung Shum, Iso-charts: stretch-driven mesh parameterization using spectral analysis, in: Proceedings of the Eurographics/ACM SIGGRAPH Symposium on Geometry Processing, 2004, pp. 45–54.

[159] Guangyu Zou, Jiaxi Hu, Xianfeng Gu, Jing Hua, Authalic parameterization of general surfaces using lie advection, IEEE Transactions on Visualization and Computer Graphics 17 (12) (2011) 2005–2014.

[160] Guangyu Zou, Jing Hua, Ming Dong, Hong Qin, Surface matching with salient keypoints in geodesic scale space, Computer Animation and Virtual Worlds 19 (3–4) (2008) 399–410.

Index

T

V

W

Printed in the United States
By Bookmasters